CALCULUS
Labs for MATLAB

Kevin M. O'Connor

JONES AND BARTLETT PUBLISHERS
Sudbury, Massachusetts
BOSTON   TORONTO   LONDON   SINGAPORE

World Headquarters
Jones and Bartlett Publishers
40 Tall Pine Drive
Sudbury, MA 01776
978-443-5000
info@jbpub.com
www.jbpub.com

Jones and Bartlett Publishers
Canada
2406 Nikanna Road
Mississauga, ON L5C 2W6
CANADA

Jones and Bartlett Publishers
International
Barb House, Barb Mews
London W6 7PA
UK

Copyright © 2005 by Jones and Bartlett Publishers, Inc.
ISBN: 0-7637-3426-8

All rights reserved. No part of the material protected by this copyright may be reproduced or utilized in any form, electronic or mechanical, including photocopying, recording, or by any information storage and retrieval system, without written permission from the copyright owner.

Production Credits
Acquisitions Editor: Timothy Anderson
Editorial Assistant: Lesley Chiller
Production Editor: Anne Spencer
Marketing Manager: Matthew Payne
Manufacturing Buyer: Therese Bräuer
Composition: NK Graphics
Text Design: Anne Spencer
Cover Design: Kristin E. Ohlin
Printing and Binding: Malloy, Inc.
Cover Printing: Malloy, Inc.
Cover Images: © Photos.com

Printed in the United States of America
09 08 07 06 05   10 9 8 7 6 5 4 3 2 1

Dedication

This book is dedicated to Gail Anderson, Stan Hargraves, Bill Audette, George Millay, Peter Gardner, Helena Zimmerman, and Ed Thomas. Few students receive as solid a mathematical upbringing as I got from these teachers, from elementary school to high school. They laid the foundation upon which this book was built.

Acknowledgments

Thanks to David Cohen and Jim Henle for their guidance in writing this book and to Anne Spencer for her help in bringing all its pieces together. Thanks also to Stephen Solomon for giving me the opportunity to write this book in the first place. And of course, special thanks to Mom and Dad, whose support over the years helped make this book possible.

Contents

1 Introducing the "Words" in MATLAB 1
- 1.1 Executing Simple Mathematical Expressions   1
- 1.2 Using Built-in Functions in MATLAB   3
- 1.3 Names for Common Constants in MATLAB   4
- 1.4 Using Variables in MATLAB   5
- 1.5 Defining Your Own Functions in MATLAB   6
- 1.6 Plotting Functions   10
- 1.7 Lists in MATLAB   14
- 1.8 Asking MATLAB for Help   17
- 1.9 Conclusion   18

2 Introducing the "Pictures" in MATLAB 19
- 2.1 Magnifying Functions   19
- 2.2 Picturing Solutions To Differential Equations in MATLAB   22
- 2.3 Plotting Slope Fields For Differential Equations   26
- 2.4 Plotting Experimental Data   29
- 2.5 Conclusion   32

3 Approximating Derivatives and Integrals 32
- 3.1 Approximating Derivatives with the Difference Quotient   32
- 3.2 Approximating Integrals with the Euler Method   37
- 3.3 Conclusion   42

4 Exact Derivatives And Integrals 43
- 4.1 Derivatives as Limits in MATLAB   43
- 4.2 Computing Derivatives with `diff`   45
- 4.3 Computing Integrals with `int`   46
- 4.4 Solving a Set of Coupled Differential Equations   49
- 4.5 Conclusion   53

5 The Theoretical Basis 54
- 5.1 Visualizing Sequence Convergence   54
- 5.2 Lower Sums and Integrals   56
- 5.3 Conclusion   60

6 Calculus Applications 61
- 6.1 Concavity and Inflection Points 61
- 6.2 Finding Minima and Maxima in MATLAB 62
- 6.3 Profiteering 65
- 6.4 Clearing the Air In MATLAB 67
- 6.5 Conclusion 70

7 Techniques of Integration 71
- 7.1 The Trapezoid Method 71
- 7.2 Simpson's Method 75
- 7.3 Vector Fields 77
- 7.4 Conclusion 79

8 Polynomial Approximations 80
- 8.1 Maclaurin and Taylor Polynomials in MATLAB 80
- 8.2 The Second-Order Euler Method 82
- 8.3 Solving Differential Equations Exactly with `dsolve` 85
- 8.4 Conclusion 89

9 Infinite Series 90
- 9.1 `symsum` and Series 90
- 9.2 Intervals and Radii of Convergence 92
- 9.3 L'Hôpital's Rule 94
- 9.4 Improper Integrals 95
- 9.5 Conclusion 98

10 The Third Dimension 99
- 10.1 Picturing Functions of Two Variables 99
- 10.2 Differentiating Functions of Two Variables 106
- 10.3 Integrating Functions of Two Variables 109
- 10.4 Conclusion 111

11 Polar, Cylindrical, and Spherical Coordinates 112
- 11.1 Using Polar Coordinates in MATLAB 112
- 11.2 Plotting in Cylindrical and Spherical Coordinates 116
- 11.3 Conclusion 121

Introduction

This text is designed to be a companion text for David W. Cohen and James M. Henle's textbook, *Calculus: The Language of Change.* While Cohen and Henle's textbook focuses mainly on solving Calculus problems using paper-and-pencil, this text teaches you how to solve Calculus problems using MATLAB, the Computer Algebra System from MathWorks, Inc. In this section, we'll briefly discuss why you might find it valuable to augment your paper-and-pencil study of Calculus with MATLAB.

One of the benefits of using MATLAB to solve Calculus problems is the immediacy with which it produces answers. As you will soon learn, solving Calculus problems can require carrying out a long sequence of computations. In the course of carrying out these computations with paper and pencil, it is very easy to lose track of the big picture—what problem are you solving, why are you solving it, and what is it teaching you. On the other hand, MATLAB automates many of the problem-solving steps for you, freeing you from worrying about the mechanics of solving the problem and allowing you to focus on the Calculus concepts that underlie the problem instead.

MATLAB is also an extremely powerful visualization tool. You can use MATLAB to visualize important concepts like the convergence of sequences, the approximation of derivatives by difference quotients, and the approximation of integrals by rectangles and trapezoids. You can also quickly and easily plot and analyze functions in two and three dimensions, including slope field plots and plots in polar, cylindrical, and spherical coordinates. A picture can truly be worth a thousand words when learning Calculus, and MATLAB makes drawing these pictures a breeze.

Lastly, you can rest assured that the time you invest in learning MATLAB now will pay off in the future, particularly if you are a mathematics, science, or engineering student. You'll be able to apply nearly all of the MATLAB skills that you'll learn in this text to your future courses and research, even if these courses have little or nothing to do with Calculus.

Hopefully, one or more of these points has helped to motivate your use of MATLAB in your study of Calculus. If so (or if your professor has made the decision for you), then it's time to begin!

CHAPTER 1
Introducing the "Words" in MATLAB

In Chapter 1 of the text, you have begun to learn how to translate between two powerful languages, English and Calculus. In this chapter, we will be begin to learn how to translate Calculus into yet a third powerful language, that of MATLAB. Fortunately, the language of MATLAB is very similar to the language of Calculus (and of mathematics in general), so knowing how to write expressions in Calculus will make learning how to write expressions in MATLAB much easier.

The goal in this chapter will be to teach you how to write basic, everyday mathematical expressions in the language of MATLAB. This will give you the basic tools you'll need for future chapters when we'll use MATLAB to tackle real, live Calculus problems.

1.1 Executing Simple Mathematical Expressions

In this section we'll show you how to express addition, subtraction, multiplication, division, and exponents in the language of MATLAB. We'll also give you some general tips for working with MATLAB.

After starting up MATLAB, you'll be presented with a blank "command window." This is where you type the expressions that MATLAB will compute, and it's also where MATLAB prints its results. To see how this works, let's have MATLAB solve the simplest of all math problems by computing the sum 1+1. Granted, asking MATLAB to compute this is a bit like asking Einstein to solve a problem from your introductory physics book, but that's OK. We have to start somewhere.

Type "1 + 1" into the MATLAB command window (don't type the quotation marks). Then, press the ENTER or RETURN key. Your command window should now look something like the following:

```
>>            1 + 1
ans =         2
```

MATLAB uses standard symbols for the basic algebraic operations of addition, subtraction, multiplication, and division. For example, the following input does a subtraction, then a multiplication, and finally a division.

```
>>          17 - 2.5
ans =       14.500

>>          7 * 4
ans =       28

>>          1/9
ans =       0.1111
```

Exponents are represented in MATLAB by the carat character, ^. For example, to compute the number google, which is ten raised to the one-hundredth power, you use the following input:

```
>>          10^100
```

After pressing ENTER or RETURN, MATLAB should produce the following output:

```
ans =       1.0000e+100
```

As you can see, MATLAB expresses numbers in scientific notation using the e character.

Note that ans always contains the result of the most recent computation. For example, we can now divide google by 100 simply by typing the following input:

```
>>          ans/100
ans =       1.0000e+098
```

Finally, you should know that you can use a semicolon to hide the output of any MATLAB input. Here's an example:

```
>>          10^100;
>>
```

MATLAB hides the output from 10^100 and simply prompts you for your next input. Using semicolons is completely optional, but it can help keep your command window from becoming cluttered.

1.2 Using Built-in Functions in MATLAB

At this stage of your mathematical career you are already familiar with a wide array of mathematical functions, including sines, logarithms, and square roots. In this section, we'll show you how to use all of these mathematical functions in MATLAB.

Names of functions in MATLAB are all more or less the same as the names you learned in math class. For example, MATLAB's sine function is named `sin`, its cosine function is named `cos`, and its tangent function is named `tan`. Thus, to compute the cosine of 3.1415926, you use the following input:

```
>>          cos(3.1415926)
ans =       -1.0000
```

Note that arguments passed into functions are always enclosed in parentheses. Also, notice that MATLAB assumes that arguments passed to trigonometric functions are in radians.

Some function names in MATLAB are a little surprising. For instance, the natural logarithm function in MATLAB is named `log`, not `ln`.

```
>>          log(2.7182818)
ans =       1.0000
```

To compute the base 10 logarithm of a number, MATLAB provides the function `log10`.

```
>>          log10(100)
ans =       2
```

Other functions that are represented only by symbols in math class are given names in MATLAB. For instance, you can take a square root by using MATLAB's `sqrt` function.

```
>>          sqrt(4)
ans =       2
```

---

1. Compute tan(3.14). After computing this value, press the up-arrow key. MATLAB will copy your last input (the one that computed tan(3.14)) back to the command line where you can edit it. Edit this copied input so that it now computes $\tan\left(\frac{3.14}{4}\right)$.

   In fact, you can access any of your previous inputs by pressing the up-arrow key multiple times.

2. `simplify` is a MATLAB function that tries to reduce a seemingly complex expression to a simpler one. As an example, `simplify` the expression $\cos^2(x) + \sin^2(x)$ to verify that it equals 1.

Note: $\cos^2(x)$ is written as `cos(x)^2` in MATLAB.

1.3 Names for Common Constants in MATLAB

In this section we'll briefly show you how to use common constants like π in your MATLAB expressions.

There are three principle constants that appear throughout mathematics and in Calculus in particular. They are none other than π, e, and the imaginary number i.

MATLAB's name for π is `pi`.

```
>>         cos(pi/4)
ans =      0.7071
```

For the imaginary number i, MATLAB actually has two different names. These are `i` and `j`.

```
>>         i * i
ans =      -1

>>         j * j
ans =      -1
```

MATLAB doesn't have a special name for e. However, you can always compute e by using the function `exp(x)`, which computes the value e^x.

```
>>         exp(1)
ans =      2.7183
```

1. Use MATLAB to show that $e^{i\pi}$ equals negative one. (The Swiss-born mathematician Leonard Euler first discovered this amazing mathematical fact in the eighteenth century.)

1.4 Using Variables in MATLAB

Most interesting mathematical expressions use variables. In this section, we'll show you how to define and use variables in MATLAB.

There are two types of variables in MATLAB:

1. Variables that have values assigned to them.
2. Variables that do not have values assigned to them.

These two types of variables are treated slightly differently in MATLAB, so you need to keep this distinction in mind.

Here is a simple example of a variable that *has* a value assigned to it:

```
>>            x = 1
x =           1
```

The variable x can now be used to assign the value 1 to another variable:

```
>>            y = x
y =           1
```

Now y is also assigned the value 1.

All this is straightforward. The subtlety comes in when you want to use a variable that has no value assigned to it. For instance, watch what happens when we try to assign t, a new variable that has no value assigned to it, to f:

```
>>            f = t
??? Undefined function or variable 't'.
```

If MATLAB could speak, it would be saying, "What the heck is t??? I've never seen it before in my life!" Before you can use the variable t, you actually need to introduce it to MATLAB with the `syms` function.

```
>>            syms t
>>            f = t
f =           t
```

Note that we don't need to use `syms` with the variable f since it has a value assigned to it, namely the variable t.

Sometimes you want to erase the value that was previously assigned to a variable. You can also use the `syms` function to do this, as the next series of inputs illustrates:

```
>>              x
x =             1

>>              syms x

>>              x
x =             x
```

After executing `syms x`, x becomes a valueless variable that is no longer assigned the value 1.

You can also completely erase a variable from your workspace with the `clear` function.

```
>>              clear x
>>              x
??? Undefined function or variable 'x'.
```

If you execute `clear` without specifying a variable name, all of the variables that are currently in use are erased. This is a one-step way to clean up when you switch from solving one problem to solving another unrelated problem.

---

1. Create a variable that is equal to the mathematical constant e.
2. Set the variable y equal to $2t^2$.
3. Erase the values from the variables you created in exercises 1 and 2.
4. Erase the variables you used in exercises 1–3.

1.5 Defining Your Own Functions in MATLAB

While studying Calculus (or practically any other mathematics or science course), you'll often be given functions to study. In this section, we'll show you how to define these functions in MATLAB. We'll also tell you about some subtleties that you should keep in mind when defining functions in MATLAB.

1.5.1 Creating Inline Functions

Let's take the following function as our example:

$$F(A, x) = A\cos(x)$$

The function F depends on x and A.

We can define this function in MATLAB using MATLAB's `inline` function, as follows:

```
>>              syms A x
>>              F = inline(char(A * cos(x)), 'A', 'x')
F =              Inline function:
                  F(A,x) = A * cos(x)
```

In the first input, `syms A x`, we introduce and erase the variables that F depends on.

In the second input, we use MATLAB's `inline` function to define F. The first argument to `inline`, `char(A * cos(x))`, is a string representation of F. We use MATLAB's `char` function to convert `A * cos(x)` into a suitable string representation. The second and third arguments to `inline`, `'A'` and `'x'`, explicitly tell `inline` that F depends on A and x. These arguments need to be enclosed in single quotations. Also, since A is listed before x, the resulting inline function is defined as `F(A,x)` rather than `F(x,A)`.

We can now use MATLAB to compute values of F in an intuitive way. For example, to compute the value of F when $A = 17$, we can simply type the following:

```
>>              F(17,x)
ans =           17 * cos(x)
```

Equivalently, we can assign the value 17 to A and then compute F.

```
>>              A = 17;
>>              F(A,x)
ans =           17 * cos(x)
```

Of course, we can also assign a value to x and compute F.

```
>>              x = 0;
>>              F(A,x)
ans =           17
```

If we later decide that x should no longer be assigned a value, we simply erase x with `syms`:

```
>>              syms x
>>              F(A,x)
ans =           17 * cos(x)
```

Note that if you use `inline` to define a function of only one variable, you don't need to explicitly tell `inline` what variables this function depends on. For instance, to create an `inline` function for

$$g(x) = x^2 + 4x,$$

you can use the following inputs:

```
>>            syms x
>>            g = inline(char(x^2 + 4*x))
g =           Inline function:
              g(x) = x^2 + 4*x
```

The `inline` function is smart enough to realize on its own that g depends on x.

1.5.2 Subtleties of Inline Functions

There are three dirty little secrets regarding inline functions that you should be aware of. First, consider the following attempt to define an inline function for $F(A, x) = A\cos(x)$:

```
>>            A = 17;
>>            F = inline(char(A * cos(x)), 'A', 'x')
F =           Inline function:
              F(A,x) = 17 * cos(x)
```

As you can see in the last line, the inline function that `inline` creates for F is completely independent of the value of A — the value of F is always computed using $A = 17$! The lesson here is that you must always erase variables with `syms` before you use them to define an inline function. You'll notice that this is exactly what we did when we defined F in the previous section.

A second subtlety regarding inline functions arises when you try to define one inline function in terms of another. For instance, in the next set of inputs we define the function `Ndiff` in terms of the function `N`:

```
>>            N = inline(char(x^2));
>>            NDiff = inline(char(N(x) - N(x - 1)));
```

When we evaluate `NDiff` at $x = 2$ we get the expected value of 3:

```
>>            NDiff(2)
ans =         3
```

However, look at what happens when we change the definition for N and then reevaluate NDiff at $x = 2$.

```
>>            N = inline(char(x^3));
>>            NDiff(2)
ans =         3
```

The value of NDiff at $x = 2$ is still 3, not 7 as it should be!

The problem is that the NDiff function continues to use the old definition of the N function. To fix this, you must explicitly redefine NDiff after redefining N:

```
>>            NDiff = inline(char(N(x) - N(x - 1)));
>>            NDiff(2)
ans =         7
```

The last subtlety worth mentioning pertains to defining constant functions in MATLAB. If you want to define a constant function such as $F = 2$, you must define this function using single quotations rather than char, as follows:

```
>>            F = inline('2')
F =           Inline function:
                  F(x) = 2
```

These three subtleties are easy to forget in the heat of doing exercises. Nevertheless, it is practically certain that you will run into them as you work through this book. Try to keep them in mind!

---

1. Create an inline function for $g(a,t) = e^{-\frac{t^2}{a}}$.
2. Use your inline function to evaluate $g(b,1)$.
3. Use your inline function to evaluate $g(4,-1)$.
4. Just as clear can be used to erase variables, it can also be used to erase inline function definitions. Use clear to erase the inline function you created for g.
5. Create an inline function for $f(x) = x^2$.
6. Create an inline function for $fsqrt(x) = \sqrt{f(x)}$.
7. Evaluate $fsqrt$ at $x = 2$.
8. Redefine f to $f(x) = 3$.
9. Evaluate $fsqrt$ at $x = 2$.
10. Update the definition of $fsqrt$ to account for the change to f.

11. Reevaluate *fsqrt* at $x = 2$.
12. Use `clear` to erase the inline functions you created for *f* and *fsqrt*.

1.6 Plotting Functions

A picture is worth a thousand words, and indeed we often gain a lot of insight into functions by plotting them. In this section, we'll present MATLAB's `ezplot` function and show you how to plot functions like our own $F(A, x)$.

Of all the MATLAB functions that do plotting, `ezplot` is the user-friendliest option of the bunch. In the next input, we use `ezplot` to plot the function $f(x) = (x - 2)^3$ for $0 \leq x \leq 4$.

```
>>              ezplot('(x-2)^3', [0, 4])
```

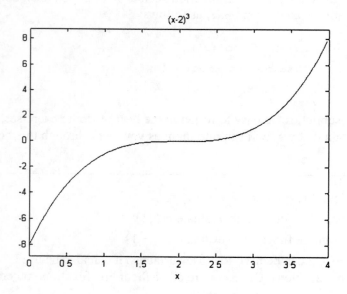

The first argument to `ezplot`, `'(x-2)^3'`, is the function to be plotted. Note that since `ezplot` requires the first argument to be a string, the single quotations around the function are essential.

The second argument, `[0, 4]`, specifies the plot domain of the independent variable. As you can see, it has the form `[xMin, xMax]`. If you want to set the plot range of the dependent variable as well, you can use a second argument of the form `[xMin, xMax, yMin, yMax]`.

There are a number of optional MATLAB functions that you can use to change the appearance of your plots. For instance, watch how the cubic plot changes as you execute the following inputs:

```
>>      xlabel('x value')
>>      ylabel('f')
>>      title('f = (x-2)^3')
>>      grid
>>      axis([1.5, 3, -0.5, 1])
```

You should see that the labels for the *x*-axis, *y*-axis, and the plot's title change after executing the first three inputs. The next command, `grid`, causes gridlines to be drawn on the plot. The last command has the form `axis([xMin, xMax, yMin, yMax])`. This redraws the current curve using the plot domain `xMin` to `xMax` and the plot range `yMin` to `yMax`.

If you prefer, you can execute related functions like these in one input. For example,

```
>>      xlabel('x value'), ylabel('f'), title('f = (x-2)^3'), grid
```

The `ezplot` function can also be used to plot inline functions. For example, in the next input we plot the inline function for $F(A,x) = A\cos(x)$:

```
>>      syms A x
>>      F = inline(char(A*cos(x)), 'A', 'x');
>>      ezplot(char(F(3,x)), [0, 4, -3, 3])
>>      grid
```

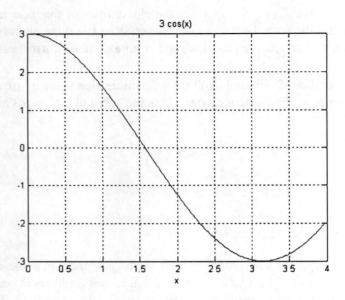

Note that since `ezplot` requires its first argument to be a string, we use `char` to convert the inline function into a string representation.

The `hold` function is the last plotting option that we will be using. The `hold` function allows you to plot multiple functions on the same set of axes, as the next set of inputs illustrates:

```
>>      ezplot('(x−2)^3', [0, 4])
>>      hold on
>>      ezplot(char(F(3,x)), [0,4,−3,3])
```

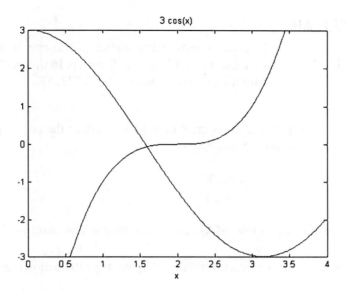

The command `hold on` tells MATLAB not to erase previous plots when drawing future plots. As you might guess, the command `hold off` causes MATLAB to revert to erasing old plots before drawing new ones.

Finally, just as defining the constant function $F = 2$ was a special case in the last section, plotting a constant function is also a special case. The following two inputs illustrate two ways to plot the constant function $F = 2$ using `ezplot`:

```
>>            ezplot('2')
>>            ezplot(F)
```

---

1. Plot the function $f_1(t) = \dfrac{1.5}{1 + t^2}$ for $-5 \leq t \leq 7$ and $-2 \leq f_1(t) \leq 2$.

2. Plot the function $f_2(t) = -\dfrac{3t}{(1 + t^2)^2}$ for $-5 \leq t \leq 7$ and $-2 \leq f_2(t) \leq 2$.

3. Use `hold` to plot f_1 and f_2 on the same set of axes. (Hint: Remember that pressing the up-arrow in the command window allows you to re-execute old inputs without having to re-type them).

4. Add gridlines to the plot of f_1 and f_2 and set the labels for the x-axis, y-axis, and the plot's title to appropriate values.

1.7 Lists in MATLAB

While the concept of a list is not a particularly mathematical one, lists of numbers are so important in MATLAB that we can't avoid bringing them up. In this section, we'll tell you what you'll need to know about lists to do Calculus in MATLAB.

1.7.1 List Basics

Let's begin by seeing how a list is created in MATLAB. In the next input, we create a list of the first five positive odd integers:

```
>>           Odds = 1:2:9
Odds =      1   3   5   7   9
```

As you can see, this input tells MATLAB to create a list of numbers from 1 to 9 in steps of two.

To find out what the first element of this list is, we type the following:

```
>>           Odds(1)
ans =       1
```

We can also find out what the second through fourth elements are by using the colon character.

```
>>           Odds(2:4)
ans =       3   5   7
```

Or we can use the colon character by itself to retrieve all of the elements.

```
>>           Odds(:)
ans =       1   3   5   7   9
```

To find out how long the list is, we use the `length` function, as follows:

```
>>           length(Odds)
ans =       5
```

We can also change the value of an element in the list. For instance, in the next input we set the last element in the list to 17.

```
>>           Odds(5) = 17
Odds =      1   3   5   7   17
```

We can also add an element to the end of the list like so:

```
>>            Odds(6) = 21
Odds =        1   3   5   7   17   21
```

1.7.2 Lists and Functions

In this book, we'll be using lists to help us compute the values of functions in two important ways.

First, suppose we want to evaluate the following function at $x = 0$, $x = 1$, $x = 2$, and $x = 3$:

$$f(x) = x^2$$

One way to do this would be to evaluate an inline function for f at each x value individually. However, there is a better way. If we "vectorize" the inline definition of f, we can evaluate f at all of these x values in one step.

In the next two inputs, we create this "vectorized" definition of f by wrapping the usual inline definition in the vectorize function.

```
>>            syms x
>>            f = vectorize(inline(char(x^2)));
```

We can now pass f a list of x values, and f will produce the corresponding list of $f(x)$ values.

```
>>            xValues = 0:1:3
xValues =     0   1   2   3

>>            fValues = f(xValues)
fValues =     0   1   4   9
```

As you can see, the ith element in the fValues list is the value of f evaluated at the ith element in the xValues list.

Of course, you can still use the vectorized f function in the usual way too.

```
>>            f(17)
ans =         289
```

We will also be using lists to evaluate recursive functions. For example, the factorial of any positive integer n, written $n!$, can be computed from the following two facts:

1. $1! = 1$
2. $n! = n*(n-1), n > 1!$

We can build a list consisting of the values of $n!$ for $1 \leq n \leq 5$ as follows. First, we create a list whose first element is the known value of the factorial function. In this case, we know that $1! = 1$.

```
>>          FactorialValues(1) = 1;
```

Then, we use a *for loop* to compute the remaining elements of the list using fact 2 above:

```
>>          for n = 2:1:5
            FactorialValues(n) = n * FactorialValues(n - 1);
            end
>>
```

When you type this *for loop* into MATLAB, hit the ENTER or RETURN key at the end of each line to begin the next one.

The first line of the *for loop* tells MATLAB what the value of n will be during each iteration of the loop. In this case, n takes on the values 2, 3, 4, and 5. The second line of the *for loop* tells MATLAB how to compute the nth value of the FactorialValues list from the $(n-1)^{th}$ element in the FactorialValues list.

After you type this *for loop* into MATLAB, you can verify that FactorialValues is now a list of the first five values of the factorial function.

```
>>          FactorialValues
FactorialValues =
            1    2    6    24    120
```

Indeed, it is!

1. Create a list, *Fib*, whose elements are the first 17 Fibonacci numbers. The nth element of the list should be the nth Fibonacci number. Recall the following:

   $Fib(1) = 1$
   $Fib(2) = 1$
   $Fib(n) = Fib(n-1) + Fib(n-2), n \geq 3$

2. What is the 17th Fibonacci number?

1.8 Asking MATLAB for Help

During your study of Calculus you'll likely come across a mathematical function that you will want to use in MATLAB but will not know how. Luckily, MATLAB provides a help system that can help you track down the information you need.

For instance, you may want to use the absolute value function, but not know MATLAB's name for it. You can begin to track down its name by typing `help` in the command window. This returns a very long list of help topics, a portion of which is printed below:

```
>>           help

             HELP topics:

             matlab\general      - General purpose commands.
             matlab\elfun        - Elementary math functions.
             matlab\specfun      - Specialized math functions.
             Etc.
```

Guessing that the absolute value function is considered an "elementary" function, we refine our search for it by typing:

```
>>           help elfun
```

This produces a list of the elementary functions, including the following:

```
             Elementary math functions.
             abs         - Absolute value.
```

There it is! We can now use `help` one last time to learn how to use the `abs` function.

```
>>    help abs

      ABS   Absolute value.
      ABS(X) is the absolute value of the elements of X.
      When X is complex, ABS(X) is the complex modulus
      (magnitude) of the elements of X.
```

It certainly took some digging, but we found what we were looking for. The more you use MATLAB's help system, the more quickly you'll be able to track down information when you need it.

---

1. Use MATLAB's help system to figure out how to compute the factorial of an integer. Compute 60 factorial. (Hint: The factorial function is a "specialized" function.)
2. Use the `help` function to learn about MATLAB's `ceil` and `floor` function. Then use these functions to compute the `ceil` and `floor` of 3.2.

1.9 Conclusion

In this brief introduction to MATLAB, we've tried to give you the basic tools you'll need to use MATLAB and to show you where to turn when you get stuck. The more you work with MATLAB, the more you'll realize just how powerful it is — and you'll be getting plenty of practice working with MATLAB in the pages to come!

CHAPTER 2

Introducing the "Pictures" in MATLAB

In Chapter 2 of the text, you've begun to see what functions and their derivatives look like. In this chapter, we'll show you how you can also use MATLAB to visualize derivatives and thereby improve your intuition about them.

2.1 Magnifying Functions

One of the most fundamental ideas from Calculus is that most functions, no matter how curvy they may appear to be, look like a straight line if you magnify them enough. In this section, you'll learn how to put functions under the magnifying glass using MATLAB.

Let's take the following function as our example:

$$f(x) = \sin(\sqrt{x}) \qquad x \geq 0$$

Seeing the sine function in $f(x)$, our first instinct is to plot this function with the plot domain $0 \leq x \leq 2\pi$.

```
>>          syms x
>>          ezplot('sin(sqrt(x))', [0, 2*pi])
```

19

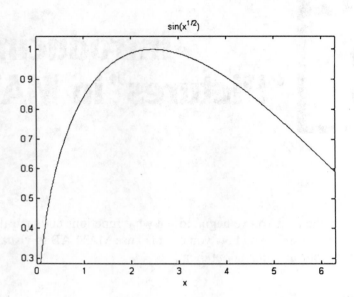

Now suppose that we want to zoom in on $f(x)$ around $x = 1.2$. All we need to do is re-plot $f(x)$ with a new plot domain that's centered on $x = 1.2$. For example, in the next plot we've used the plot domain $0.7 \leq x \leq 1.9$.

```
>>              ezplot('sin(sqrt(x))', [0.7, 1.9])
```

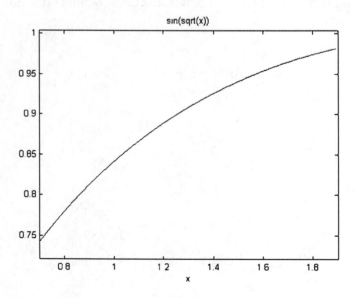

Of course, you can continue to shrink the plot domain of x to further magnify the curve. Here we continue to magnify $f(x)$ around the point $x = 1.2$ by further shrinking the plot domain of x to $1.15 \leq x \leq 1.25$.

```
>>                  ezplot('sin(sqrt(x))', [1.15, 1.25])
```

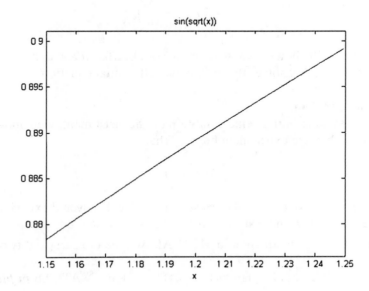

One note before we move on. If you have been typing each of the above inputs from scratch each time, you don't need to do this! You can simply press the up-arrow key, edit your previous input, and re-execute it. The plot of $f(x)$ will be redrawn.

The following are some rather curvy functions. Plot them in MATLAB using the given plot domains and identify the x value at which $f(x)$ looks to be the curviest. Then zoom in on $f(x)$ around this x value until $f(x)$ looks linear.

1. $f(x) = x^2$ $-2 \leq x \leq 2$
2. $f(x) = \sqrt{x}$ $0 \leq x \leq 4$
3. $f(x) = \sin(x)$ $-\pi \leq x \leq \pi$
4. $f(x) = e^{-x^2}$ $-5 \leq x \leq 5$
5. $f(x) = x\ln(x)$ $0.1 \leq x \leq 3$

2.2 Picturing Solutions to Differential Equations in MATLAB

In Section 2.4 of the text, you were introduced to differential equations. Then, in Section 2.6, you learned that the rate of growth of yeast cells could be modelled by the following differential equation in particular:

$$Y' = kY\left(1 - \frac{Y}{c}\right)$$

In this section, we'll show you how to use a MATLAB *m-file* and the `ode45` function to plot the function, $Y(t)$, that solves this particular differential equation.

2.2.1 Setting Up the Problem

Let's suppose that the initial value condition on the yeast mentioned above is that there is one yeast colony when the experiment begins. That is,

$$Y(0) = 1.$$

Let's also suppose that for our particular yeast cells, we've discovered experimentally that the values of the constants k and c are $k = 0.6$ and $c = 700$.

To solve this differential equation in MATLAB, we need to carry out two steps:

1. Define the differential equation $Y' = kY(1 - \frac{Y}{c})$ in a MATLAB *m-file*
2. Feed this *m-file* to MATLAB's differential equation solver, `ode45`

In the next two sections, we'll show you how to do both of these steps. Then we'll show you how to plot the solution that `ode45` finds and evaluate it at specific times.

2.2.2 Creating an *m-file* Function for the Differential Equation

An *m-file* function, like an inline function, is a way to define a function in MATLAB. *m-file* functions are particularly useful for defining differential equations, so we'll be using them whenever we need to solve differential equations.

We will now create an *m-file* function, `yeastDiffEq(t,Y)`, that computes the value of Y' at any point (t, Y) according to the differential equation

$$Y' = kY\left(1 - \frac{Y}{c}\right).$$

To create this *m-file* function, you first need to locate the MATLAB program's menu bar. The menu bar contains standard drop down menus like *File*, *Edit*, and *View*. From the

File menu, select *New* and then *M-file*. A window for a text editor should pop up. Now, type the following definition of the `yeastDiffEq` function into the text editor (don't type the line numbers):

```
1        function dYdt = yeastDiffEq(t,Y)
2              k = 0.6;
3              c = 700;
4              dYdt = k*Y*(1-Y/c);
```

Once you've finished typing this *m-file*, save it by selecting *Save As* from the text editor's *File* menu. If it's not already, you should name this *m-file* `yeastDiffEq.m`. An *m-file* function must have the same name as the function that it defines, with `.m` tacked on the end of the function's name.

Let's examine the pieces of this *m-file* function.

In line 1, we begin the *m-file* function with the keyword `function`. This tells MATLAB that we are creating an *m-file* function rather than some other type of *m-file*.

In the next part of line 1, we tell MATLAB that the value of the function `yeastDiffEq` will be stored in the variable `dYdt`. Note that you can name this variable anything you like. In this case, the name `dYdt` is particularly apt since the value of the `yeastDiffEq` function is the value of Y'.

In the final part of line 1, we tell MATLAB that `yeastDiffEq` is the name of the *m-file* function we are defining. We also tell MATLAB that `yeastDiffEq`'s first argument is t and its second argument is Y. Note that when you are defining an *m-file* function for a differential equation, the independent variable must be the function's first argument and the dependent variable must be the second argument. Also note that the arguments of the *m-file* function must be both the independent variable and the dependent variable, even if the differential equation is independent of one or both of these variables. For instance, even though Y' doesn't explicitly depend on t, t must still be an argument of the `yeastDiffEq` function.

In the remaining lines of the *m-file* function, we set the values of k and c, compute the value of Y' according to $Y' = kY(1 - \frac{Y}{c})$, and store the result in `dYdt`, as we promised to do in line 1.

Having defined the differential equation in `yeastDiffEq.m`, we can now use MATLAB's `ode45` function to solve the differential equation.

2.2.3 Solving the Differential Equation with `ode45`

Once you've written an *m-file* function for your differential equation, the hard part is done. All that's left to do is feed this *m-file* function to `ode45`:

```
>>         [t, Y] = ode45('yeastDiffEq', 0:1:20, 1);
```

The first argument to `ode45`, `'yeastDiffEq'`, is the name of the *m-file* function that defines the differential equation to be solved. This name should be contained within single quotation marks.

The second argument to `ode45` has the form `tMin:tStep:tMax`. This is a list of the t values at which we want to evaluate the solution function, $Y(t)$. Since we are interested in the first twenty hours of the yeast experiment, we are asking `ode45` to compute the values of $Y(t)$ from 0 hours to 20 hours in steps of one hour.

The third and final argument to `ode45` specifies the initial value condition for the differential equation. This is the value of $Y(t)$ when t equals `tMin`. In the case of our yeast experiment, we know that $Y(0) = 1$, so we set the third argument to 1.

The output of `ode45` is a pair of lists, `t` and `Y`, that represent the solution to the differential equation. `t` is the list of times at which `ode45` computed the value of the solution function $Y(t)$, and `Y` is the corresponding list of these computed values. For example, consider the following:

```
>>          t(10)
ans =       9
>>          Y(10)
ans =       168.4554
```

What this is telling us is that $Y(9) = 168.4554$. That is, there were about 168 yeast colonies nine hours into the experiment.

2.2.4 Plotting and Evaluating the Solution to the Differential Equation

While the format that `ode45` uses to represent the solution function $Y(t)$ may seem a little strange to you, it's commonly used by MATLAB. In fact, it's so common that MATLAB provides a function, `plot`, that plots functions that are represented in exactly this way.

```
>>          plot(t,Y)
```

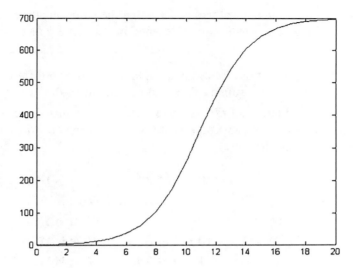

As you can see, the first argument to `plot` is the list of values for the plot's independent variable. The second argument is the corresponding list of values for the plot's dependent variable.

If you want to evaluate $Y(t)$ at a particular value of t, this is a little tricky. For example, let's suppose that we want to know how many yeast colonies there are seven hours into the experiment. That is, we want to evaluate $Y(t)$ at $t = 7$. We first need to determine the index of the element in the `t` list that equals 7. In this case, this is pretty easy since the index into the `t` list is always one more than the value of t that it picks out. In general, trial and error may be the quickest way to determine this!

```
>>              t(8)
ans  =          7
```

We now know that the eighth element in the `t` list corresponds to $t = 7$. Thus, the eighth element in the `Y` list contains the value of Y at $t = 7$.

```
>>              Y(8)
ans  =          60.9773
```

Thus, there are approximately 61 yeast colonies after the first seven hours of the experiment.

Clearly, this can be a tedious process. Fortunately, we are often more concerned with the qualitative shapes of solutions to differential equations than with their quantitative values. And certainly on a qualitative level you can see that the solution curve for this differential equation represents logistic growth, as you learned in Section 2.6 of the textbook.

Use `ode45` and `plot` to plot the solutions, $Y(t)$, to the following differential equations over the given domain for t. Also, evaluate $Y(t)$ at the given value(s) of t.

Note: Once you've written an *m-file* for the differential equation in exercise 1, you can reuse it for the remaining exercises. Just modify it to represent the new differential equation and resave it. Be sure to resave it!

1. $Y' = t$ $Y(0) = 2$ $0 \leq t \leq 3$ What is $Y(2)$?
2. $Y' = -t$ $Y(0) = 2$ $0 \leq t \leq 3$ What is $Y(2)$?
3. $Y' = Y$ $Y(0) = 1$ $0 \leq t \leq 3$ What is $Y(2)$?
4. $Y' = -Y$ $Y(0) = 1$ $0 \leq t \leq 3$ What is $Y(2)$?
5. $Y' = \frac{1}{t}$ $Y(1) = 0$ $1 \leq t \leq 20$ What is $Y(20)$?
6. $Y' = \frac{1}{Y}$ $Y(1) = 1$ $1 \leq t \leq 30$ What is $Y(4)$?
7. $Y' = \frac{1}{2Y}$ $Y(1) = 1$ $1 \leq t \leq 30$ What is $Y(4)$? $Y(9)$? $Y(25)$?
8. $Y' = t + Y$ $Y(0) = 1$ $0 \leq t \leq 3$ What is $Y(0)$?
                                                    (What had it *better* be?)
9. $Y' = t - Y$ $Y(0) = 1$ $0 \leq t \leq 3$ What is $Y(3)$?
10. $Y' = tY$ $Y(0) = 2$ $0 \leq t \leq 3$ What is $Y(2)$?

In exercise 7, can you guess what the function $Y(t)$ is, based on $Y(4)$, $Y(9)$, and $Y(25)$?

2.3 Plotting Slope Fields For Differential Equations

Slope fields are another very useful way of picturing differential equations. In this section, we'll show you how to plot slope fields for differential equations in MATLAB. We'll then show you how to plot specific solutions to a differential equation through its slope field.

2.3.1 Plotting a Slope Field for Differential Equation

Let's create the slope field for the following differential equation:

$$y' = t - y$$

We can plot the slope field for this differential equation in MATLAB using the following inputs:

```
>>          [T, Y] = meshgrid(-1:1:5, -2:1:4);
>>          syms t y
>>          dy = vectorize(inline(char(t - y), 't', 'y'));
>>          arrowY = dy(T, Y);
>>          arrowT = ones(size(Y));
>>          quiver(T, Y, arrowT, arrowY);
```

After executing the last input, you should see the following slope field:

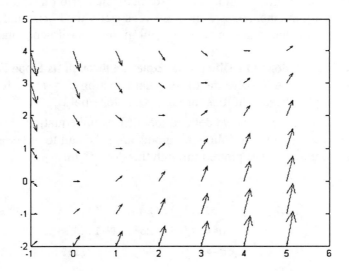

As you would expect, the slope of the arrow plotted at the point (t,y) is equal to the derivative of the solution function $y(t)$ at this same point.

Let's look at what each of the MATLAB inputs above is doing, especially the inputs that use the new MATLAB functions `meshgrid` and `quiver`.

The first input uses MATLAB's `meshgrid` function to create the two-dimensional grid of points on which the slope arrows will be plotted. The first argument to `meshgrid`, -1:1:5, is the list of t values at which the slope arrows will be drawn. The second argument to `meshgrid`, `-2:1:4`, is the list of *y* values at which the slope arrows will be drawn. Thus, slope arrows are plotted on a square grid bounded by the points $(-1,-2)$ and $(5,4)$, with each slope arrow separated from its nearest neighbors by one unit in both the horizontal and vertical directions. This two-dimensional grid of points created by `meshgrid` is stored in the lists `T` and `Y`.

The following two inputs define the inline function, `dy`, which represents the differential equation $y' = t - y$. This function is vectorized so that we can easily evaluate it at all of the points in the two-dimensional grid defined by `T` and `Y`.

The next two inputs specify the vertical and horizontal components of each slope arrow that is drawn. The *y*-component of the arrow drawn at the point (t,y) is set to the value of y' at this point using the function `dy`. The *t*-component of this and every arrow is set to one using MATLAB's `ones` function. Thus, the slope of the arrow drawn at any point (t,y) is $\frac{y'}{1}$, or simply y', as it should be for a slope field.

Finally, the last input uses the MATLAB function `quiver`[1] to plot the slope arrows. The first two arguments, `T` and `Y`, define the two-dimensional grid on which the arrows are plotted. The last two arguments, `arrowT` and `arrowY`, specify the horizontal and vertical components of the arrow that is drawn at each point in this two-dimensional grid.

2.3.2 Plotting Specific Solutions to a Differential Equation through Its Slope Field

You can now use `ode45` to solve the differential equation $y' = t - y$ for various initial value conditions and plot these solutions through your slope field.

For instance, let's suppose you've defined the differential equation $y' = t - y$ in the *m-file* `diffEq.m`. If you execute the following inputs after the call to `quiver` above, you will see three different solution curves plotted through the slope field.

```
>>          hold on
>>          [t, y] = ode45('diffEq', [-1:.1:5], [2]);    plot(t,y)
>>          [t, y] = ode45('diffEq', [-1:.1:5], [0]);    plot(t,y)
>>          [t, y] = ode45('diffEq', [-1:.1:5], [-1.8]); plot(t,y)
>>          hold off
```

---

[1] A quiver, as you may know, is a case for holding arrows. It seems the creators of MATLAB were not without a sense of humor.

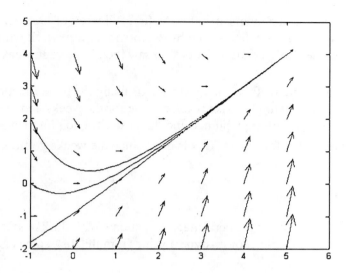

It's easy to see that the slopes of these solutions are in perfect agreement with the slope field.

1–10. Plot the slope fields for the differential equations in Section 2.2's exercises. Then plot the specific solution you found for each differential equation on the same set of axes as the slope field.

Note: For exercise 5, plot the slope field for $t > 0$. For exercises 6 and 7, plot the slope fields for $Y > 0$.

2.4 Plotting Experimental Data

In this, the final section of the chapter, we'll show you how to plot data that you collect during an experiment. In this particular case, we'll be using data that was collected during the flu season of 2003–2004.

The following is the actual data that the CDC collected during the 2003–2004 flu season:

Week	0	1	2	3	4	5	6	7	8	9	10	11	12
% Flu Cases	0.9	1.0	1.3	1.3	1.8	2.4	2.7	3.7	4.5	5.4	7.4	7.5	7.6

Week	13	14	15	16	17	18	19	20	21	22	23	24	25
% Flu Cases	5.2	2.9	2.1	1.9	1.8	1.5	1.4	1.3	1.1	1.0	1.0	1.1	0.9

Weeks are numbered such that week 0 is the first full week of October 2003. "% Flu Cases" is the percentage of all patients seen by doctors in a given week who had flu-like symptoms. For example, 7.6% of all patients who saw their doctors in week 12 had symptoms that seemed flu-like.

We can make a picture of this data in MATLAB using the `plot` function that we introduced in Section 2.2. We'll simply make two lists, one for the weeks and one for the percentage of patients with flu-like symptoms, and then use `plot` to plot them.

First we create the list for weeks. This is easy, since the weeks go from 0 to 25 in steps of one.

```
>>          weeks = 0:1:25;
```

Next, we create the list for the percentage of patients with flu-like symptoms. It would be very easy to add, delete, or mistype a value from this list, so be careful when you type it in.

```
>>          percentFluCases = [0.9, 1.0, 1.3, 1.3, 1.8, 2.4, 2.7,
                3.7, 4.5, 5.4, 7.4, 7.5, 7.6, 5.2, 2.9, 2.1, 1.9,
                1.8, 1.5, 1.4, 1.3, 1.1, 1.0, 1.0, 1.1, 0.9];
```

We now have the values for the independent variable in `weeks` and the corresponding values for the dependent variable in `percentFluCases`. We can now use `plot` to make a picture of this data. We'll also label the picture it creates.

```
>>          plot(weeks, percentFluCases)
>>          xlabel('Weeks')
>>          ylabel('% Patients with Flu-like Symptoms')
>>          title('The 2003-2004 Flu Season')
>>          grid
```

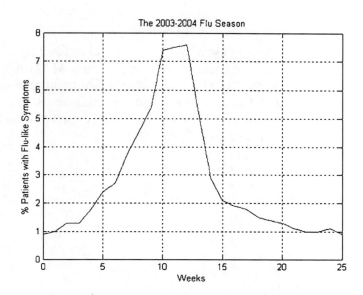

Voilà! Using MATLAB we have created an information-packed picture of the 2003–2004 flu season.

---

1. The following is the CDC data from the 2002–2003 flu season, a much more typical flu season with regards to its timing and severity:

Week	0	1	2	3	4	5	6	7	8	9	10	11	12
% Flu Cases	1.2	1.2	1.1	1.3	1.3	1.4	1.4	1.3	1.4	1.3	1.3	1.6	2.0

Week	13	14	15	16	17	18	19	20	21	22	23	24	25
% Flu Cases	1.8	1.5	1.8	2.4	2.8	3.2	3.0	2.9	2.3	1.9	1.7	1.5	1.2

Using MATLAB, make a plot of this data just as we did above. Then, using `hold`, plot the 2003–2004 data on the same axes as the 2002–2003 data. How did the 2002–2003 flu season differ from the 2003–2004 flu season?

2.5 Conclusion

In this chapter we learned how to use MATLAB to visualize functions and their derivatives. For instance, we learned how to zoom in on functions using MATLAB's `ezplot` function. We also learned how to use MATLAB to plot solutions to differential equations using `ode45` and `plot`. We then moved on to slope fields, learning how they could be plotted with the `quiver` function. Lastly, we learned how `plot` could be used to visualize experimental data, such as the flu season data from 2003–2004.

CHAPTER 3

Approximating Derivatives and Integrals

In this chapter, we'll begin by using MATLAB to approximate derivatives with the difference quotient. We'll then implement the Euler method in MATLAB and use it to approximate the values of definite integrals.

3.1 Approximating Derivatives with the Difference Quotient

In Sections 3.1 and 3.2 of the text, you learned that you could approximate the derivative of a function, $f(t)$, by computing the difference quotient.

$$\frac{df}{dt} = f'(t) \approx \frac{f(t + \Delta t) - f(t)}{\Delta t}$$

In this section, we'll use MATLAB to visualize how well the difference quotient approximates $f'(t)$ as Δt approaches 0.

We'll use the difference quotient to approximate the derivative of the function:

$$f(t) = t^3.$$

We first create a vectorized inline function for $f(t)$:

```
>>          syms t
>>          f = vectorize(inline(char(t^3)));
```

Now we use $f(t)$ to define a new inline function, `approxDer(t,deltat)`, that approximates the derivative of $f(t)$ using the difference quotient. We `vectorize` this function as well.

```
>>          syms deltat
>>          approxDer = vectorize(inline(char((f(t+deltat)-f(t))/
                       deltat), 't', 'deltat'));
```

We can now watch what happens to the difference quotient's approximation of $f'(t)$ by plotting it for various values of Δt. For comparison, we'll plot both $f(t)$ and the approximation of $f'(t)$ on the same set of axes, using a dashed line for $f(t)$ and a solid line for the approximation of $f'(t)$.

We start with $\Delta t = 5$, a very large value.

```
>>      deltat = 5;
>>      T = -5:0.1:5;
>>      plot(T, approxDer(T, deltat))
>>      hold on
>>      plot(T, f(T), '-.')
>>      grid, axis([-5, 5, -25, 40]), hold off
```

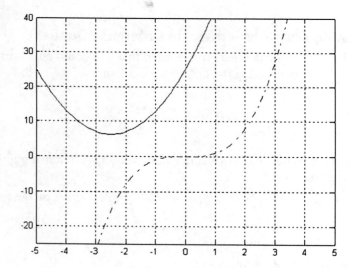

To plot $f(t)$ with a dashed line, you'll notice that we've used "$-.$" as an optional third argument to plot. Other possibilities include:

'.-'	Plot with connected dots
'x-'	Plot with connected x's
'.'	Plot with disconnected dots
'x'	Plot with disconnected x's

Note that you cannot control the plotting style in this manner when you use `ezplot`.

Now take a second and ask yourself how well the solid curve approximates the derivative of $f(t)$, the dashed curve. Are there any glaring discrepancies? One obvious problem is that the rate of change of the dashed curve at $t = 0$ is 0, yet the value of the supposed derivative function at $t = 0$ is greater than 20 — a far cry from 0!

Let's see if the derivative approximation gets any better when we shrink Δt to 1.

```
>>    deltat = 1;
>>    plot(T, approxDer(T, deltat))
>>    hold on
>>    plot(T, f(T), '-.')
>>    grid, axis([-5, 5, -25, 40]), hold off
```

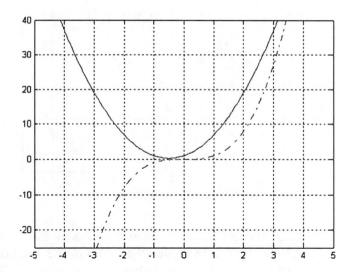

Yes, this is definitely a better approximation of the derivative. The approximated derivative function's value at $t = 0$ is now about 1, much closer to the correct value of 0.

Let's do this one more time, now with $\Delta t = 0.1$. This should produce an excellent approximation of the derivative function.

```
>>      deltat = 0.1;
>>      plot(T, approxDer(T, deltat))
>>      hold on
>>      plot(T, f(T), '-.')
>>      grid, axis([-5, 5, -25, 40]), hold off
```

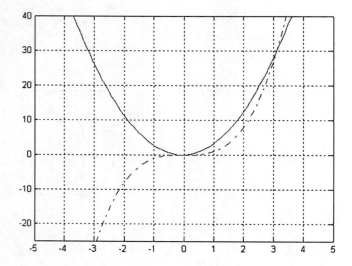

And indeed it does. The values of the approximated derivative function are large at t values where the rate of change of $f(t)$ is large, and the values of the approximated derivative function are small at t values where the rate of change of $f(t)$ is small. This is exactly what we expect from a derivative function.

Show that the difference quotient also leads to good approximations of the derivatives of the following functions. Try some Δt values that are large and some that are small and note how the quality of the derivative approximation is affected.

Note: Don't forget the subtleties of using inline functions that we discussed in Section 1.5.2. Specifically, you will need to redefine the `approxDer` function each time you redefine the `f` function. You should also execute `syms deltat` before you redefine `approxDer`.

1. $f(t) = (t-1)(t+3)t^2$
2. $f(t) = 2t^2$
3. $f(t) = 4t$
4. $f(t) = \dfrac{1}{t}$ (Note: Use a plot domain of $-2 \leq t \leq 2$)
5. $f(t) = \dfrac{1}{t^2 + 1}$
6. $f(t) = \sin(t)$
7. $f(t) = \cos(t)$
8. $f(t) = e^t$
9. $f(t) = \ln(t)$

3.2 Approximating Integrals with the Euler Method

In this section, we'll use the Euler method to approximate the value of the integral

$$\int_0^3 f'(t)\,dt = f(3) - f(0),$$

where

$$f'(t) = \tfrac{1.5}{1 + t^2}$$

and

$$f(0) = 0.$$

We'll also plot the Euler method's approximation of $f(t)$ from $t = 0$ and $t = 3$.

3.2.1 Review of the Euler Method

As you know, the Euler method uses $f'(t)$ to approximate values of $f(t)$ in a step-by-step manner. For instance, if we know the value of f at time $t = t_0$, then we can approximate the value of f at time $t = t_0 + \Delta t$ to be:

$$f(t_0 + \Delta t) \approx f(t_0) + f'(t_0)\Delta t$$

We'll use this fact to implement the Euler method in MATLAB and thereby approximate the integral

$$\int_0^3 f'(t)dt.$$

3.2.2 Implementing the Euler Method in MATLAB

We begin our MATLAB implementation of the Euler method by defining some useful constants, including the limits of integration, the value of Δt, and the initial value of f.

```
>>      tI = 0;
>>      tF = 3;
>>      deltat = 0.6;
>>      fInitialValue = 0;
```

Next, we create the list of times at which we'll approximate the value of $f(t)$ using the Euler method.

```
>>           times = tI:deltat:tF
times =      0   0.6000   1.2000   1.8000   2.4000   3.0000
```

We are now ready to create a list of the Euler method approximations of f at each of these times. We'll call this list fEuler1, and we'll build it recursively starting with the fact that the initial value of f is 0. Note that dfdt is the list of values of f' evaluated at each time in times.

```
>>      fEuler1(1) = fInitialValue;
>>      for i = 1:1:(length(times)-1)
            dfdt(i) = 1.5/(1 + times(i)^2);
            fEuler1(i+1) = fEuler1(i) + dfdt(i)*deltat;
        end
```

As you can see, we're computing fEuler1(i+1) from fEuler1(i) in exactly the manner prescribed by the Euler method, namely:

$$f(t_0 + \Delta t) \approx f(t_0) + f'(t_0)\Delta t.$$

Thus, after executing this *for loop*, the ith element in the `fEuler1` list is the Euler method's approximation of f at the ith time in the `times` list. The next set of inputs illustrates this.

```
>>              times
times =         0   0.6000   1.2000   1.8000   2.4000   3.0000
>>              fEuler1
fEuler1 =       0   0.9000   1.5618   1.9306   2.1429   2.2760
```

In particular, we see that the Euler approximation of $f(3)$ equals 2.2760. Thus, we can now approximate the integral of $f'(t)$ from $t = 0$ to $t = 3$, as follows:

$$\int_0^3 \frac{1.5}{1+t^2}\, dt = f(3) - f(0) \approx 2.2760$$

We can also plot the values in `fEuler1` to see what the Euler method's approximation of $f(t)$ looks like for $0 \le t \le 3$.

```
>>              plot(times, fEuler1)
```

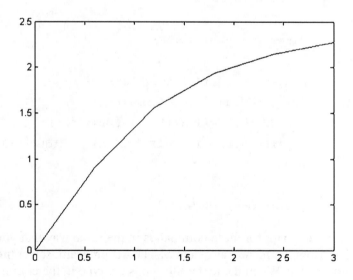

You may be wondering how well this curve approximates $f(t)$. One way to get a sense of this is to make several different plots of the Euler method's approximation of $f(t)$ with smaller and smaller values of Δt. For instance, how does the approximation of $f(t)$ change when $\Delta t = 0.2$ and $\Delta t = 0.1$?

We'll now show you an easy way to try out different values of Δt in the Euler Method using an *m-file* script.

3.2.3 Using a Script to Automate the Euler Method in MATLAB

An *m-file* script is nothing but a set of MATLAB commands collected into a single *m-file*. The beauty of a script, however, is that you can execute all of the commands in the script simply by typing the script's name in the command window.

The following is an *m-file* script that contains all of the commands we used in the previous section to implement the Euler method.

```
1           syms fEuler1 dfdt
2
3           tI = 0;
4           tF = 3;
5           deltat = 0.6;
6           fInitialValue = 0;
7
8           times = tI:deltat:tF;
9
10          fEuler1(1) = fInitialValue;
11          for i = 1:1:(length(times)-1)
12              dfdt(i) = 1.5/(1 + times(i)^2);
13              fEuler1(i+1) = fEuler1(i) + dfdt(i)*deltat;
14          end
15
16          plot(times, fEuler1)
```

You create an *m-file* script for these commands in the same way that you create an *m-file* function. First, from the menu bar of the MATLAB program, select the *File* menu, then *New*, and finally *M-file*. When the text editor pops up, type in the commands above (leaving out the line numbers of course!). Finally, save this *m-file* file, using any name you like. In our case, we named the script `euler1Script.m`. When you return to the command window, you should find that you can now execute all of the commands in `euler1Script.m` simply by typing `euler1Script`!

For instance, in the next set of inputs we call `euler1Script` three different times to generate the Euler method's plots for Δt values of 0.6, 0.2, and 0.1. Note that before each call to `euler1Script`, we are manually changing the value of `deltat` in `euler1Script.m` and re-saving it.

```
>>    euler1Script
>>    hold on
>>    euler1Script
>>    euler1Script
>>    hold off
```

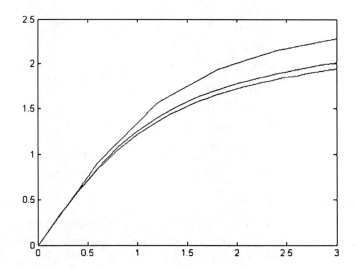

There are two things worth noting about `euler1Script.m` before we move on. First, you'll notice that we used the command

```
syms fEuler1 dfdt
```

in the first line of `euler1Script.m`. This ensures that the `fEuler1` and `dfdt` lists are erased before we start using them in the script.

Second, you may wonder why we're using the number 1 in the name of the `fEuler1` list. Let's just say that Chapter 3 is not the last time you will encounter the Euler method! In fact, it will come back bigger and better in Chapter 8, so keep your Euler method scripts in a safe place.

For each of the following functions and initial value conditions, use our Euler method script to approximate the integral

$$\int_0^3 f'(t)\,dt$$

Also, plot the Euler method approximation of $f(t)$ for $0 \le t \le 3$. Try using different values for Δt and note how the approximations change.

1. $f'(t) = 2$ $f(0) = 2$
2. $f'(t) = 2t$ $f(0) = 2$
3. $f'(t) = 2t^2$ $f(0) = 2$
4. $f'(t) = 2t^3$ $f(0) = 2$
5. $f'(t) = \cos(t)$ $f(0) = 1$
6. $f'(t) = \cos(2t)$ $f(0) = 1$
7. $f'(t) = \sin(t)$ $f(0) = 0$
8. $f'(t) = \sin(2t)$ $f(0) = 0$
9. $f'(t) = e^t$ $f(0) = 1$
10. $f'(t) = e^{-t^2}$ $f(0) = 1$

3.3 Conclusion

In this chapter, you used MATLAB to approximate derivatives with the difference quotient, and you used MATLAB to approximate definite integrals with the Euler method. In the next chapter, we'll show you how to use MATLAB to compute these derivatives and integrals exactly.

CHAPTER 4

Exact Derivatives and Integrals

In Chapter 4 of the textbook, you learned how to find exact derivatives and integrals of a number of different functions. In this chapter, we'll show you how to compute the derivatives and integrals of all these functions and more in MATLAB.

4.1 Derivatives as Limits in MATLAB

As you've learned in the text, the derivative of a function is defined as the following limit:

$$f'(x) = \lim_{x \to 0} \frac{f(x + \Delta x) - f(x)}{\Delta x}$$

In this section, we'll show you how to use MATLAB's `limit` function to compute derivatives using this definition.

Let's take the following function as our example:

$$f(x) = 2x + x^4$$

First, we define f in MATLAB.

```
>>          syms x
>>          f = inline(char(2*x + x^4));
```

Now, we use MATLAB's `limit` function and the definition of a derivative to compute the derivative of f:

```
>>          syms dx
>>          limit( (f(x+dx)-f(x))/dx, dx, 0)
ans =       2+4*x^3
```

The `limit` function requires three arguments. The first argument, `(f(x + dx) - f(x))/dx`, is the expression of which we want to take the limit. The second and third arguments, `dx` and `0`, tell MATLAB which limit we are taking. Together, they can be read "as dx approaches 0." As you can see, taking this particular limit does indeed produce the derivative of $f(x)$, namely $f'(x) = 2 + 4x^3$.

Here's another example. This time we'll find the derivative of the following function, in which n is a constant:

$$g(x) = x^n$$

```
>>              syms x n dx
>>              g = inline(char(x^n), 'x', 'n');
>>              limit( (g(x+dx,n)-g(x,n))/dx, dx, 0)
ans =
                x^n/x*n
```

This answer is indeed equal to $g'(x)$, but it can be simplified further. In the next input, we use MATLAB's `simplify` function to do the simplification for us.

```
>>              simplify(ans)
ans =
                x^(-1+n)*n
```

This is probably closer to how you would write this derivative.

Using `limit` and the definition of a derivative, compute the derivative, $f'(x)$, of each of the following functions. Note that n represents a constant.

1. $f(x) = x^n$
2. $f(x) = (nx)^n$
3. $f(x) = \sin(x)$
4. $f(x) = \sin(nx)$
5. $f(x) = \cos(x)$
6. $f(x) = \sin(nx)$
7. $f(x) = e^x$
8. $f(x) = e^{nx}$
9. $f(x) = \ln(x)$
10. $f(x) = \ln(nx)$

Considering your answers to problems 1–8, does there appear to be a general relationship between the derivative of the function $f(x)$ and the derivative of $f(nx)$? Does this relationship hold up for problems 9–10?

4.2 Computing Derivatives with `diff`

While you certainly could compute any derivative with `limit`, MATLAB also provides a function that computes derivatives of functions directly. This function is called `diff`.

```
>>          diff(2*x + x^4, x)
ans =       2+4*x^3
```

Pretty easy, huh? In fact, since we've already defined $f(x) = 2x + x^4$ as the inline function `f(x)` in MATLAB, we could just as easily have computed this derivative like so:

```
>>          diff(f(x), x)
ans =       2+4*x^3
```

We can also use `diff` to create an inline function, `dfdx`, which represents the derivative function $f'(x)$:

```
>>          dfdx = inline(char(diff(f(x), x)))
dfdx =      Inline function:
                dfdx(x) = 2+4*x^3
```

Now we can use `dfdx` to evaluate the derivative at various values of x. For instance, to compute the value of the derivative of $f'(x)$ at $x = 17$, we would input:

```
>>          dfdx(17)
ans =       19654
```

Of course, you can also plot `dfdx` with `ezplot` as you would any other inline function. Try it out!

In the next series of exercises, you will use `diff` and the following five pairs of functions to test out the product rule, the quotient rule, and the chain rule of differentiation.

Note: You may find it easier to show two expressions as equal if you `simplify` each of them first.

1. $f(x) = (x + 1)(x - 3)$ $g(x) = x^3 + 2x + 1$
2. $f(x) = \sqrt{x + 3}$ $g(x) = 3x^2 - x + 2$
3. $f(x) = \cos(x)$ $g(x) = \dfrac{1}{1 + x^2}$
4. $f(x) = \tan(x)$ $g(x) = e^{-x^2}$
5. $f(x) = \ln(x)$ $g(x) = e^x$

First, let's test out the product rule for differentiation. For each pair of functions above, carry out the following steps:

    a. Define the function $h(x) = f(x)g(x)$.
    b. Use `diff` to create functions for $f'(x)$, $g'(x)$, and $h'(x)$.
    c. Compute $f'(x)g(x) + f(x)g'(x)$ and show that it equals $h'(x)$.

Next, let's test the quotient rule for differentiation. Using the functions $f(x)$ and $g(x)$ above, carry out the following steps:

    a. Define the function $h(x) = \dfrac{f(x)}{g(x)}$.
    b. Use `diff` to create functions for $f'(x)$, $g'(x)$, and $h'(x)$.
    c. Compute $\dfrac{g(x)f'(x) - g'(x)f(x)}{(g(x))^2}$, and show that it equals $h'(x)$.

Finally, test the chain rule for differentiation. Again, use the functions for $f(x)$ and $g(x)$ above to carry out the following steps:

    a. Define the function $h(x) = g(f(x))$.
    b. Use `diff` to create functions for $f'(x)$, $g'(x)$, and $h'(x)$.
    c. Compute $g'(f(x))f'(x)$ and show that it equals $h'(x)$.

4.3 Computing Integrals with `int`

Finding the anti-derivative, or integral, of a function is also quite easy in MATLAB, thanks to MATLAB's `int` function.

For instance, to compute the definite integral

$$\int_{1}^{20} \frac{1}{s}\, ds$$

Exact Derivatives and Integrals

we type the following:

```
>>              syms s
>>              int(1/s, s, 1, 20)
ans =           2*log(2)+log(5)
```

MATLAB not only provides the answer, it provides the exact answer!

As you can see, the first argument to `int`, `1/s`, is the function we want to integrate. The variable we are integrating is in respect to the second argument, `s`. The third and fourth arguments, `1` and `20`, respectively, are the limits of integration.

To convert `int`'s result to a decimal value, you can use MATLAB's `double` function.

```
>>              double(ans)
ans =           2.9957
```

The function `int` can also handle indefinite integrals, as the following input illustrates:

```
>>              int(1/s, s)
ans =           log(s)
```

By eliminating the limits of integration, we're telling `int` to solve the indefinite integral. The astute reader will note, however, that something is missing from MATLAB's answer. That something is a constant of integration, which should be included in the solution of any indefinite integral. MATLAB assumes that you know this, however, and omits it.

Chances are that `int` will be able to integrate just about any function you throw at it. There are, however, some integrals that no one (not even mighty MATLAB) knows how to compute exactly. For instance, look at what happens when we ask MATLAB to integrate $f(x) = x^x$ as a definite and as an indefinite integral.

```
>>              int(x^x, x, 1, 2)
ans =           int(x^x, x = 1 .. 2)
>>              int(x^x, x)
ans =           int(x^x, x)
```

When MATLAB spits your input back at you, MATLAB is saying, "I don't know how to compute that." All is not lost, however. We can at least find a numerical approximation for the definite integral by wrapping the call to `int` in MATLAB's `double` function:

```
>>              double(int(x^x, x, 1, 2))
ans =           2.0504
```

`double` instructs `int` to approximate the value of the definite integral when `int` can't compute the value exactly.

Quite frankly, it's very unlikely that you'll run across many integrals that can't be handled by `int`, but `double` is always ready to rumble should `int` ever get stuck.

Use `int` to compute the following definite integrals. If `int` "doesn't know" how to integrate a particular definite integral, use `double` to approximate the integral.

1. $\int_{0}^{1.5} x^3 dx$

2. $\int_{0}^{2\pi} \sin(x) dx$

3. $\int_{1}^{2} \frac{dx}{x}$

4. $\int_{-5}^{5} e^{-x^2} dx$

You'll notice that MATLAB's answer to exercise 4 involves a strange function named `erf`. If you're curious, you can type `help erf` on the command line to learn more about this function. However, if you're interested only in the numeric value of the integral, you can compute this wrapping the `int` call in `double`.

Now use `int` to compute the following indefinite integrals. Don't forget that MATLAB does not include the constant of integration when it solves an indefinite integral, but you can be sure that your Calculus professor will expect to see it on exams!

Note that n represents a constant.

1. $\int dx$

2. $\int x^n dx$

3. $\int nx^n dx$

4. $\int \sin(x) dx$

5. $\int n\sin(x)dx$

6. $\int (\sin(x) + x^n)dx$

7. $\int \frac{1}{x}dx$

8. $\int \frac{n}{x}dx$

9. $\int \left(\frac{n}{x} - x^n\right)dx$

10. $\int \sec^2(x)dx$

11. $\int n\sec^2(x)dx$

Considering your answers to these problems, propose how the indefinite integral of $f'(x)$ is related to the indefinite integral of $nf'(x)$. Also, write an expression for the integral of $f(x) + g(x)$ in terms of the integral of $f(x)$ and the integral of $g(x)$.

4.4 Solving a Set of Coupled Differential Equations

We've already seen how an *m-file* function and `ode45` can be used to solve a single, uncoupled differential equation like $y' = y$. In this section, we'll show you how to use these same tools to solve a set of coupled differential equations.

As a motivating example, consider what you learned in Section 3.8 of the textbook. You learned that the spread of an infectious disease in a population of horses could be modelled by the following set of coupled differential equations:

$$S' = -aSI$$
$$I' = aSI - bI$$
$$R' = bI$$

$S(t)$ is the number of horses susceptible to the disease at time t, $I(t)$ is the number of horses infected at time t, and $R(t)$ is the number of horses that have either recovered or died from the disease at time t.

We will now use MATLAB to solve this set of coupled differential equations under the following conditions:

$$S(1) = 2180$$
$$I(1) = 1$$
$$R(1) = 0$$
$$a = 0.001$$
$$b = 0.2$$

4.4.1 Creating an *m-file* Function for the Differential Equations

In order to use `ode45` to solve these differential equations, we express them in the following *m-file* function named `epidemic.m`:

```
1                   function diffEqs = epidemic(t, solutions)
2
3                   diffEqs = zeros(size(solutions));
4
5                   a = 0.001;
6                   b = 0.2;
7
8                   S = solutions(1);
9                   I = solutions(2);
10                  R = solutions(3);
11
12                  diffEqs(1) = -a*S*I;
13                  diffEqs(2) = a*S*I - b*I;
14                  diffEqs(3) = b*I;
```

Think of `diffEqs(1)`, `diffEqs(2)`, and `diffEqs(3)` as synonyms for S', I', R' respectively. Thus, for example, we set `diffEqs(1)` equal to `-a*S*I`, the expression for S'.

Similarly, in lines 8–10, we create the variables `S`, `I`, and `R` as shorthand names for `solutions(1)`, `solutions(2)`, and `solutions(3)`, respectively. This allows us to succinctly express the differential equations in lines 12–14 in terms of `S`, `I`, and `R`.

Note that it doesn't matter how you assign `diffEqs(1)`, `diffEqs(2)`, and `diffEqs(3)` to S', I', and R', nor does it matter how you assign `solutions(1)`, `solutions(2)`, and `solutions(3)` to S, I, and R. What *does* matter is that the function represented by `diffEqs(i)` is the derivative of the function represented by `solutions(i)`.

Exact Derivatives and Integrals

As for the remaining lines in `epidemic.m`, the only novelty is the use of `zeros` in line 3. This MATLAB function initializes `diffEqs` before it is used in the *m-file*, and it is a required part of any *m-file* function that defines a set of coupled differential equations.

4.4.2 Solving the Differential Equations with `ode45`

We can now use `ode45` to solve the set of coupled differential equations defined in `epidemic.m`.

```
>>          [t,solutionFuncs] = ode45('epidemic', 1:0.5:50, [2180, 1, 0]);
```

As you can see, the call to `ode45` looks pretty much the same whether we're solving a set of differential equations or a single differential equation. The first argument, `'epidemic'`, is still the name of the *m-file* function that defines the differential equations, and it is still contained in single quotations. The second argument, `1:0.5:50`, still specifies the set of t values at which the solution functions will be evaluated. In this case, we are asking `ode45` to compute solutions from $t = 1$ to $t = 50$ in time steps of 0.5.

The third argument, `[2180, 1, 0]`, is the only argument that is slightly different. It now specifies the three initial values $S(1)$, $I(1)$, and $R(1)$, respectively. In general, the ith value in this list of initial values should be the initial value of the function represented by `solutions(i)` in the *m-file* function. For example, since `solutions(1)` represents the function S, the first value in the list `[2180, 1, 0]` is the initial value for S.

We now turn our attention to the output of `ode45`. The `t` output is simply the list of times from $t = 1$ to $t = 50$ at which the solution functions are evaluated. For instance, in the next input we learn that the thirteenth element of the `t` list corresponds to $t = 7$:

```
>>          t(13)
ans =       7
```

The `solutionFuncs` output from `ode45` consists of the values of the solution functions $S(t)$, $I(t)$, and $R(t)$, evaluated at each of the times in the `t` list. For example, here is the value of S at $t = 7$:

```
>>          solutionFuncs(13,1)
ans =       56.1356
```

Here is the value of I at $t = 7$:

```
>>          solutionFuncs(13,2)
ans =       1.3932e+003
```

And here is the value of R at $t = 7$:

```
>>          solutionFuncs(13,3)
ans =       731.6219
```

As you can see, `solutionFuncs` is indexed in the same way that `solutions` is indexed in the *m-file*. Thus, in our case, S values are accessed using index 1, I values are accessed using index 2, and R values are accessed using index 3. In general, to access the ith value of the jth solution function, you type `solutionFuncs(i,j)`.

Of course, we can plot the curves for $S(t)$, $I(t)$, and $R(t)$ by passing the output from `ode45` to `plot`.

```
>>          plot(t, solutionFuncs(:,1)), hold on
>>          plot(t, solutionFuncs(:,2), '.-')
>>          plot(t, solutionFuncs(:,3), 'x-'), hold off
```

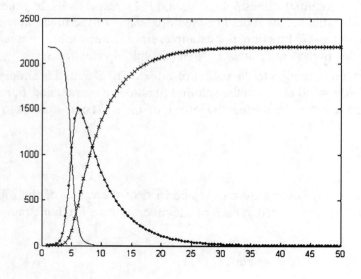

As you can see, we've plotted $S(t)$ using a solid curve, $I(t)$ using connected dots, and $R(t)$ using connected x's.

Your goal now is to solve Assignment #1 and Assignment #2 in Section 3.8 of the textbook using `ode45`. You've got all the tools you'll need to solve this set of coupled differential equations and others like it!

4.5 Conclusion

In this chapter we have covered the lion's share of MATLAB's Calculus capabilities. Specifically, you learned how to use MATLAB to take limits, compute derivatives, and calculate integrals using `limit`, `diff`, `int`, and `double`. You also learned how to solve a set of coupled differential equations using *m-file* functions and `ode45`.

Hopefully, after working through this chapter you are feeling much more comfortable doing Calculus problems in MATLAB. If not, never fear — you'll be getting a lot more practice in the chapters to come!

CHAPTER 5: The Theoretical Basis

How far we've come! We began this book by learning how to solve 1+1 in MATLAB, and now we're using it to solve complicated systems of differential equations. Not bad for only four chapters of work.

In fact, you now know more than enough MATLAB to solve a wide variety of problems in Calculus. Therefore, our emphasis for the remainder of this book will be on using what you already know about MATLAB to sharpen your Calculus skills and intuition. Of course, we'll also introduce new MATLAB functions as they become relevant to what you're learning in the textbook.

Chapter 5 in the textbook, which builds up the theory needed to prove the Fundamental Theorem of Calculus, is a good place to make this transition. After all, computers are rarely used to prove mathematical theorems, but they are more and more often used to provide the intuition that inspires these proofs.

5.1 Visualizing Sequence Convergence

In Chapter 5 of the text, you learned about the convergence of sequences. One way to get a sense for whether a sequence converges is to plot it. For instance, in the next set of inputs we plot the first twenty values of the sequence $a_n = \frac{n+2}{n}$.

```
>>      syms n
>>      a = vectorize(inline(char((n+2)/n)));
>>      N = 1:1:20;
>>      A = a(N);
>>      plot(N, A, 'x')
```

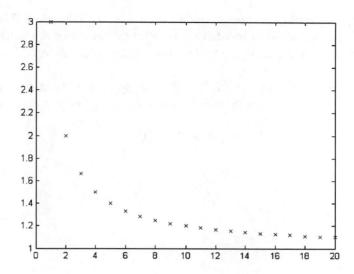

It's pretty clear from this plot that the sequence is converging to a value in the neighborhood of one. In fact, if you plot more sequence values, you can probably convince yourself that this sequence is converging to exactly one.

Plot each of the following sequences as above. If you think the sequence converges, to what value do you think it converges?

1. $a_n = \frac{n + 201}{n}$

2. $b_n = \frac{n}{n + 201}$

3. $c_n = \frac{n^2}{2000000n}$

4. $d_n = \cos((.99)^n)$
5. $e_n = \sin((-n)^n)$
6. $f_n = e^{(-.9)^n}$

7. $g_n = 3^{\frac{n}{2}}$

8. $h_n = 3^{\frac{-n}{2}}$
9. $i_n = \ln(\frac{1}{n})$
10. $j_n = (1 + \frac{1}{n})^n$

Using what you now know about the convergence of the sequences above, do the following sequences converge? If so, to what value? Plot the sequence to confirm your answer. Be sure to use Proposition 5.5 from the text in formulating your answers!

Note: In the following exercises, *a*, *b*, *c*, etc. are the same sequences defined in the previous exercise. For instance, the terms of sequence *j* are defined in exercise 10 above:

$$j_n = (1 + \tfrac{1}{n})^n$$

1. $5a$
2. $12c - 2b$
3. dj
4. $\dfrac{e}{j}$
5. $i + j$

5.2 Lower Sums and Integrals

In Section 5.7 of the text, you were shown how to use lower sums to compute the exact area under the curve $y(x) = x^2$ from $x = 0$ to $x = 1$. In this section, we'll use the same lower sums approach to compute the exact area under the curve $y(x) = x^3$ from $x = 0$ to $x = 1$. In other words, we'll use lower sums to compute the exact value of the following integral:

$$\int_0^1 x^3 dx$$

Of course, MATLAB will do most of the work for us!

Let's begin by presenting the MATLAB script that we'll be using to visualize and compute the lower sums. We'll call this script `lowerSums.m`.

```
1       syms x
2       y = inline(char(x^3));
3
4       xMin = 0; xMax = 1;
5       yMin = 0; yMax = 1;
6       n = 10;
7
8       dx = (xMax-xMin)/n;
9       rectangleSum = 0;
```

```
10
11              hold off
12              xValues = xMin:dx:(xMax-dx);
13              for i = 1:1:length(xValues)
14                  x = xValues(i);
15                  fill([x x (x+dx) (x+dx)], [0 y(x) y(x) 0], 5);
16                  rectangleSum = rectangleSum + y(x)*dx;
17                  hold on
18              end
19
20              syms x
21              ezplot(char(y(x)), [xMin, xMax, yMin, yMax]);
22              hold off
23
24              rectangleSum
```

After saving this *m-file* as lowerSums.m and typing lowerSums in the command window, a plot and a number will be produced.

```
>>              lowerSums
```

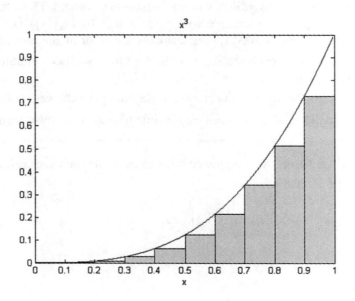

```
rectangleSum = 0.2025
```

As you can see, these MATLAB inputs generate a picture of the lower sum for $y(x) = x^3$ that is produced by $n = 10$ equal-width rectangles from $x = 0$ to $x = 1$. The inputs also compute the sum of the areas of these rectangles, 0.2025, which is a lower sum for the following integral:

$$\int_0^1 x^3 dx$$

Let's examine the `lowerSums` script and see how it works.

In lines 1–9, we define the function $y(x)$ and the variables used in the script. `xMin` and `xMax` specify the limits of integration, and `yMin` and `yMax` specify the plot range of the plot that is created. `n` is the number of rectangles to be drawn, and `dx` is the width of each of these rectangles. `rectangleSum` will hold the the sum of the areas of these rectangles.

In lines 11–18, we use a *for loop* to draw each rectangle using MATLAB's `fill` function. The first argument of `fill`,

```
[x x (x+dx) (x+dx)]
```

specifies the x-coordinates of the corners of the rectangle. The second argument,

```
[0 y(x) y(x) 0]
```

specifies the corresponding y-coordinates of the rectangle's corners. Thus, together, the first two arguments of `fill` specify a rectangle with corners at $(x,0)$, $(x,y(x))$, $(x + dx, y(x))$, and $(x + dx, 0)$. The third argument of `fill`, 5, determines the color of the rectangle.

Meanwhile, as we draw each rectangle in the *for loop*, we also compute its area and add it to `rectangleSum` in line 16.

Finally, in lines 20–24, we plot the curve for $y(x)$ and print the value of `rectangleSum`.

Now that we've digested `lowerSums.m`, it's time to play with lower sums a bit.

You will now use MATLAB and lower sums to compute the exact value of the following integral:

$$\int_0^1 x^3 dx$$

1. Increase the value of n in `lowerSums.m`. Does the lower sum approach the value you expect for the integral? If so, for what value of n do you come within 1% of the exact value of the integral?

2. Recall that in Section 5.7 of the textbook you were able to compute a general expression for the area of the n rectangles under the curve $y = x^2$ for $0 \leq x \leq 1$. Can you find a similar general expression for the areas of the n rectangles under the curve $y(x) = x^3$ from $x = 0$ to $x = 1$? This is a lower sum for our integral.

   Note that MATLAB is quite adept at doing symbolic arithmetic. For example, MATLAB can give you a closed expression for the sum of the first n positive integers:

   ```
   >>              syms n k
   >>              symsum(k, 1, n)
   ans =           1/2*(n+1)^2-1/2*n-1/2
   ```

MATLAB can also give you a closed expression for the sum of the first n cubes, although you may want to `simplify` the answer it gives you.

3. What is the limit as n approaches infinity of the lower sum you just computed for the integral:
$$\int_0^1 x^3 dx$$
   Does this limit equal the actual value of the integral?

4. Suppose we change the lower limit of integration in this integral from 0 to -0.5. Will the MATLAB script we used in this section compute a proper lower sum for this new integral? Try it and see.

5. For any function $y(x)$, would the MATLAB script we used in this section compute a proper lower sum for the following integral:
$$\int_0^1 y(x) dx$$

   If you think the answer is yes, explain why. If you think the answer is no, use our MATLAB script to find a function for which it does not compute a proper lower sum.

6. Modify the MATLAB script `lowerSums.m` to compute lower sums for the following integral:

$$\int_0^1 x^4 \, dx$$

7. Answer questions 1–4 above for this new integral.

5.3 Conclusion

In this chapter, we began by using MATLAB to visualize the convergence and divergence of sequences. We then used MATLAB to visualize and compute lower sums for the following integrals:

$$\int_0^1 x^3 \, dx \quad \text{and} \quad \int_0^1 x^4 \, dx$$

Finally, we used MATLAB to show that these lower sums converged to the exact values of both of these integrals.

In the next chapter, we'll change gears a bit and show you how to use MATLAB to study the critical points of functions.

CHAPTER 6

Calculus Applications

In this chapter, we'll use MATLAB to study the critical points of functions. We'll also introduce you to a very interesting problem regarding the destruction of ozone in the stratosphere.

6.1 Concavity and Inflection Points

We'll begin this chapter with some exercises involving concavity and inflection points. Since you already know all the MATLAB you need to do these exercises, we'll let you get right to them.

Plot each of the following curves and identify all of the inflection points by eye.

1. $y(x) = 5x - 1 + x^2$
2. $y(x) = x^3 - 2x^2 - 2x + 1$
3. $y(x) = 88x^8 + 8x^2 + 8$
4. $y(x) = \dfrac{1}{1 + x^4}$
5. $y(x) = x\sqrt{x} - 2x$
6. $y(x) = x + 3\sin\left(\dfrac{2x}{3}\right)$
7. $y(x) = x^2 e^x$
8. $y(x) = x - e^x$
9. $y(x) = (\ln(x))^2 + x$
10. $y(x) = x^x + x$

6.2 Finding Minima and Maxima in MATLAB

MATLAB has two functions, `fminsearch` and `fminbnd`, for computing the minima and maxima of curves. In this section we'll show how to use these functions.

Throughout this section, we'll use the following curve as our example:

$$y(x) = 0.1x^4 - 1.1x^3 + 3x^2 - x + 5$$

This curve has two local minima and one local maximum.

```
>>      syms x
>>      y = inline(char(0.1*x^4 - 1.1*x^3 + 3*x^2 - x + 5));
>>      ezplot(char(y(x)), [-5, 10, -10, 25])
>>      grid
```

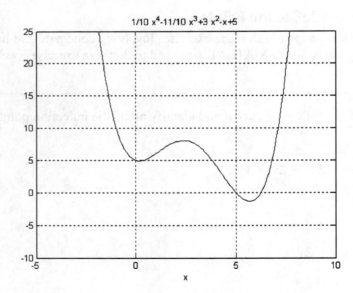

6.2.1 Finding the Local Minima

To find the local minima of $y(x)$ we use fminsearch like so:

```
>>                 [xMin, yMin] = fminsearch(char(y(x)), 0)
xMin =             0.1851
yMin =             4.9108
```

As you can see, the first argument to fminsearch, char(y(x)), is a string representing $y(x)$. The second argument, 0, specifies that fminsearch should begin searching for the minimum of $y(x)$ at $x = 0$.

The output of fminsearch is essentially the ordered pair (x_{min}, y_{min}). Note, however, that even though $y(x)$ has two minima, fminsearch finds only one of them. To find the second local minimum, we need to execute fminsearch again with a search value closer to the second minimum. When we try $x = 5$, we find the second local minima.

```
>>                 [xMin, yMin] = fminsearch(char(y(x)), 5)
xMin =             5.6918
yMin =             -1.3825
```

The function fminbnd works just like fminsearch, except that you can tell it to search for local minima within a specific domain of x values. For instance, to search for a local minimum of $y(x)$ between $x = 4$ and $x = 8$, you would use fmindbnd like so:

```
>>                 [xMin, yMin] = fminbnd(char(y(x)), 4, 8)
xMin =             5.6918
yMin =             -1.3825
```

6.2.2 Finding the Local Maxima

To find the local maximum of $y(x)$, we use the fact that the local maxima of any function $f(x)$ are the local minima of the function $-f(x)$. Thus, we can also use fminsearch to find the local maximum of $y(x)$. The following input illustrates this:

```
>>                 [xMin, yMin] = fminsearch(char(-y(x)), 2)
xMin =             2.3731
yMin =             -7.9924
```

Thus, the value of the local maximum of $y(x)$ is +7.9924 and occurs at $x = 2.3731$.

6.2.3 Nuances of `fminsearch`

Depending on the curve, `fminsearch` can be very sensitive to the initial search value. For example, look at what happens when we start the search for the local maximum of $y(x)$ at $x = 0$ instead of $x = 2$ as we did above.

```
>>            [xMin, yMin] = fminsearch(char(-y(x)), 0)

              Exiting: Maximum number of function evaluations has
              been exceeded
xMin =        -3.1691e+026
yMin =        -1.0087e+105
```

Wow. The function `fminsearch` went searching for the maximum to the left of $x = 0$ instead of to the right and got quite lost.

Indeed, it's a good idea to verify that the points returned by `fminsearch` are actually local minima and local maxima. The derivative tests for critical points are an easy way to check.

For example, here's a quick way to check that the local maximum that `fminsearch` found at $x = 2.3731$ is in fact a local maximum. We first define inline functions for the first and second derivatives of $y(x)$.

```
>>            dydx = inline(char(diff(y(x), x)));
>>            dydx2 = inline(char(diff(dydx(x), x)));
```

We now evaluate these derivatives at the point $x = 2.3731$.

```
>>            dydx(2.3731)
ans =         5.1498e-005
>>            dydx2(2.3731)
ans           -2.9045
```

The first derivative of $y(x)$ at $x = 2.3731$ is as close to 0 as you can expect from a computer, and the second derivative of $y(x)$ at $x = 2.3731$ is clearly negative. As you learned in Section 6.7 of the text, this means that $y(x)$ does indeed have a local maximum at $x = 2.3731$.

For each of the following functions, first find and classify the relative extrema *without using* MATLAB. Then, use MATLAB to confirm your calculations by plotting the function and using `fminsearch` and/or `fminbnd` to find its extrema.

1. $y(x) = -2x^2 + 17x - 3$
2. $y(x) = -x^3 + 17x^2 - 3x + 1$
3. $y(x) = \dfrac{\sqrt{x}}{1 + x^2}$ How does MATLAB identify the point at $x = 0$?
4. $y(x) = \sin(x)e^x$ Find only the extrema for $0 \leq x \leq 3\pi$
5. $y(x) = x - \sin^2(x)$ Find only the extrema for $-\pi \leq x \leq \pi$

In light of exercise 5, explain how `fminsearch` handles inflection points.

6.3 Profiteering

In Section 6.12 of the text, you learned how to maximize profit using demand functions. In this section, we'll continue in this vein and investigate a new demand function.

As you know from Section 6.12 of the text, a demand function tells us the demand for a product given the product's price. For instance, in Section 6.12, you were trying to sell a product with the following demand function:

$$n(p) = 360{,}000 - 20{,}000p$$

This demand function tells you that if the price of your product is $10, then there will be a demand for $n(10) = 160{,}000$ of your products. As the price of this product increases, the demand for it decreases, as can be seen from a plot of the demand function.

```
>>    ezplot('360000-20000*p', [0, 18])
>>    xlabel('price')
>>    ylabel('demand (# of units)')
>>    title('Demand vs. Price')
```

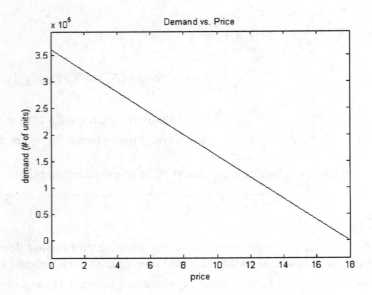

As you can see, the demand for the product decreases at a constant rate as its price increases. When the price reaches about $18, there is no demand for the product whatsoever.

With this as background, you are now ready to tackle the next set of exercises.

In the following set of exercises, we'll imagine that you have a new product you want to sell. Naturally wanting to make the most profit off your product as you can, you go through the following steps.

1. Your crack staff of marketers informs you that the demand function for your product is:

$$n(p) = 1000e^{-\frac{(p-100)^2}{250}} + 0.0001p^3 - 0.0000005p^4 + 100$$

Plot the demand function. Then give your marketers a raise — they were up all last night working on that demand function!

2. There's a reason your marketers are working so hard — they're excited about your product. In fact, they claim that your product is so wonderful that there'll be demand for it no matter what its price is!

Knowing that marketing types are prone to exaggerate, examine the demand function for yourself. To what extent does it support the marketers' claim? Is there a particular price for which demand is the greatest?

3. The folks in production tell you that it will cost $90 per unit to make your product. What is the total cost of producing enough product to meet demand when the product's price is p?
4. What would be your total profit from selling p products? Plot the profit curve.
5. What price maximizes your profit? Is there another price that comes in a close second?
6. Oh no! You just talked to production, and they regret to inform you that they underestimated the production cost for your product! Now they're "pretty sure" production costs will be *exactly* $93.51 per unit and not a penny more. Replot the profit curve using this new production cost.
7. What price maximizes your profit now?
8. You should discover that you can charge two quite different prices and still make the same profit. How do you explain this, given that demand for your product is so much higher around $110 than it is around $180? What price would you charge? Why?
9. Suppose production comes back and tells you that production costs have unexpectedly risen to $100 per unit. After firing everyone on your production team, plot the profit curve and find the new price that maximizes your profit.
10. Why is this new profit-maximizing price not surprising, even in light of the extremely high demand that exists when the product costs about $100?

6.4 Clearing the Air in MATLAB

In Section 6.15 of the text, you were introduced to a very cool problem involving the destruction of ozone in the stratosphere. You have all the MATLAB tools you need to solve it, now all you need is the motivation. Hopefully, that's what we'll provide for you in this section.

The careful application of Calculus to the well-known problem of ozone destruction leads to the following set of differential equations:

$$O'_3 = bOO_2 - O_3$$
$$O'_2 = -aO_2 - bOO_2 + cO_3$$
$$O' = 2aO_2 - bOO_2 + cO_3$$

$O_3(t)$, $O_2(t)$, and $O(t)$ are the number of ozone molecules, oxygen molecules, and oxygen atoms in the stratosphere at a particular time t. The constant a controls the rate at which molecular oxygen, O_2, breaks down into atomic oxygen, O. The constant b controls the rate at which molecular oxygen reacts with atomic oxygen to form ozone, O_3. Finally, the constant c controls the rate at which ozone breaks back down into molecular oxygen and atomic oxygen. For more details about any of these equations or constants, you should refer to the textbook.

Now for the questions: What happens to the number of ozone molecules when they react with molecular and atomic oxygen according to the equations above? Will the ozone break down completely into molecular and atomic oxygen? Or might the ozone levels increase without bound? Or perhaps over time the number of ozone molecules remains more or less constant.

Here is a MATLAB *m-file*, ozone.m, that will start you on the path to answering these questions.

```
1       function diffEqs = ozone(t, solutions)
2
3       diffEqs = zeros(size(solutions));
4
5       a = 1;
6       b = 1;
7       c = 1;
8
9       ozone = solutions(1);
10      mol   = solutions(2);
11      atom  = solutions(3);
12
13      diffEqs(1) = b*atom*mol - ozone;
14      diffEqs(2) = -a*mol - b*atom*mol + c*ozone;
15      diffEqs(3) = 2*a*mol - b*atom*mol + c*ozone;
```

And here are the MATLAB inputs you'll need to find and plot the solutions to the differential equations defined in this *m-file*.

```
>>      [t, y] = ode45('ozone', [0:.1:5], [10 10 10]);
>>      plot(t,y(:,1)), hold on
>>      plot(t,y(:,2), '.-')
>>      plot(t,y(:,3), 'x-'), hold off
```

After executing these inputs, you should see the following plot. The solid curve represents ozone, the curve plotted with dots represents molecular oxygen, and the curve plotted with x's represents atomic oxygen.

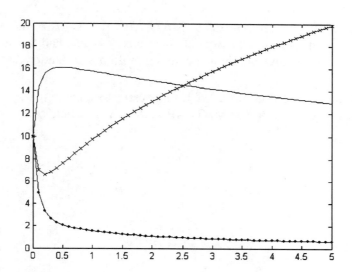

Interestingly, this plot suggests that all of the oxygen in the stratosphere will eventually be in the form of atomic oxygen. It also suggests that molecular oxygen, the same old stuff we breathe, is highly unstable in the stratosphere since the number of O_2 molecules drops rapidly. Meanwhile, the number of ozone molecules remains relatively constant but does tend to decrease over time.

The values we've used above are quite generic — all the constants are equal to 1 and all the initial values are set to 10. Play with these a bit and see how the solutions change. Are there any other solutions that look qualitatively different than the ones plotted above? If so, how are they different?

Now, consider the two additional reactions that are also presented in Section 6.15.

$$O + NO_2 \rightarrow NO + O_2$$
$$NO + O_3 \rightarrow NO_2 + O_2$$

Create differential equations for these reactions and add them to `ozone.m`. How does adding nitrogen dioxide to the equation affect the levels of ozone in the stratosphere?

There is much more information about this problem in the textbook, so if the question has piqued your interest, check out Section 6.15 for all the details. It even contains a reference to another textbook from which these ideas and equations were taken.

6.5 Conclusion

In this chapter, you learned how to use MATLAB's `fminsearch` and `fminbnd` functions to find the local minima and maxima of functions. You also had the opportunity to solve some interesting Calculus problems involving profit maximization and ozone destruction.

In the next chapter, we'll use MATLAB to investigate some new ways to approximate integrals. We'll also learn how to use MATLAB to draw vector fields.

CHAPTER 7

Techniques of Integration

In this chapter, we'll begin by investigating the rectangle, trapezoid, and Simpson methods for approximating integrals. We'll then learn how to plot vector fields in MATLAB.

7.1 The Trapezoid Method

In Chapter 5 we used MATLAB to approximate the area under the curve $y(x)$ using a set of n equal-width rectangles. As you now know, we were actually implementing the rectangle method for approximating integrals. In this section, we'll convert our implementation of the rectangle method into an implementation of the trapezoid method. We'll then see how well these two methods approximate the value of several integrals.

Let's begin by recalling how we implemented the rectangle method to approximate the following integral:

$$\int_0^1 x^3 dx$$

Here is the script we used in Chapter 5, which we will now call `rectangleScript.m`.

```
1               syms x
2               y = inline(char(x^3));
3
4               xMin = 0; xMax = 1;
5               yMin = 0; yMax = 1;
6               n = 10;
7
8               dx = (xMax-xMin)/n;
9               rectangleSum = 0;
```

```
10
11          hold off
12          xValues = xMin:dx:(xMax-dx);
13          for i = 1:1:length(xValues)
14              x = xValues(i);
15              fill([x x (x+dx) (x+dx)], [0 y(x) y(x) 0], 5);
16              rectangleSum = rectangleSum + y(x)*dx;
17              hold on
18          end
19
20          syms x
21          ezplot(char(y(x)), [xMin, xMax, yMin, yMax]);
22          hold off
23
24          rectangleSum
```

After running this script, we generated the following picture and numerical approximation of the integral:

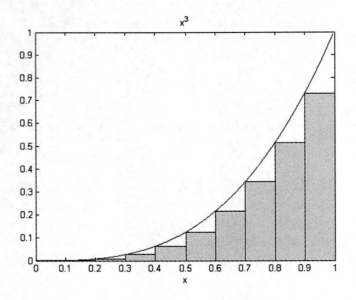

```
rectangleSum = 0.2025
```

To convert this script into a script that implements the trapezoid method, we only need to make two changes. First, we need to draw trapezoids rather than rectangles. Second, we need to compute `trapezoidSum` rather than `rectangleSum`.

To draw trapezoids instead of rectangles, we need to change the second $y(x)$ to $y(x + dx)$ in `fill`'s list of y-coordinates. Thus, the call to `fill` for the trapezoid method is:

```
15          fill([x x (x + dx) (x + dx)], [0 y(x) y(x + dx) 0], 5);
```

Next, to compute the area of these trapezoids, we use the fact that each trapezoid's area is:

$$\tfrac{1}{2}(y(l) + y(r))dx$$

In this equation, l is the left x-coordinate of the trapezoid and r is the right x-coordinate of the trapezoid. This tells us exactly how to modify line 16 to compute `trapezoidSum`:

```
16          trapezoidSum = trapezoidSum + 1/2 * (y(x) + y(x+dx)) * dx;
```

Now replace any remaining references to `rectangleSum` with `trapezoidSum`, and our implementation of the trapezoid method will be complete. Save it as `trapezoidScript.m`.

We can now use the trapezoid method to approximate the following integral:

$$\int_0^1 x^3 dx$$

Our result using three trapezoids is provided below for reference.

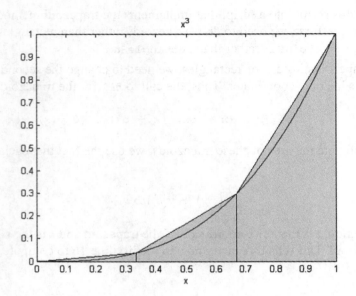

```
trapezoidSum = 0.2778
```

Now it's time to use the trapezoid and rectangle methods to approximate some integrals.

Approximate each of the following integrals using the rectangle and trapezoid methods with $n = 6$ partitions. Compare their results to each other and to the exact value of the integral. Does either approximation method compute the integral exactly?

1. $\int_0^{10} x\,dx$

2. $\int_0^{10} x^2\,dx$

3. $\int_0^{10} x^3\,dx$

4. $\int_0^{10} x^4\,dx$

5. $\displaystyle\int_0^9 \sqrt{x+1}\,dx$

6. $\displaystyle\int_1^2 e^{x-2}\,dx$

7. $\displaystyle\int_1^5 \ln(x)\,dx$

8. $\displaystyle\int_0^\pi \sin(x)\,dx$

9. $\displaystyle\int_0^\pi \sin(2x)\,dx$

10. $\displaystyle\int_1^3 x^x\,dx$

7.2 Simpson's Method

In Section 7.6 of the text you learned about a third way to approximate the value of an integral, called Simpson's method. Rendering Simpson's method in MATLAB would be rather tricky since it isn't trivial to draw all the approximating parabolas. However, you can easily modify your scripts for the rectangle and trapezoid methods to compute the Simpson's method approximation of the integral:

$$\int_{x_{min}}^{x_{max}} y(x)\,dx$$

In fact, carrying out this modification is the first problem in the next set of exercises.

Using your scripts for the rectangle method and trapezoid method as templates, write a script that computes `simpsonSum`. Since `simpsonSum` is the Simpson's method analogue of `rectangleSum` and `trapezoidSum`, it should compute the area under the approximating parabolas for the curve `y(x)` between `xMin` and `xMax`.

Hints:

    a. `n` is still the number of partitions, and `dx` is still the width of each partition. Assume that `n` is an even number as required by Simpson's method.

    b. `simpsonSum` should be a function of `dx`, `y(i)`, `y(i+dx)`, and `y(i+2dx)`.

    c. Recall that the area under one approximating parabola can be written as:

$$\left(\tfrac{1}{3}y(x) + \tfrac{4}{3}y(x+dx) + \tfrac{1}{3}y(x+2dx)\right)dx$$

Now you can approximate the value of most any definite integral using the rectangle method, trapezoid method, or Simpson's method.

In the following exercises, approximate the given definite integral using these three methods with $n = 6$ partitions. Then, compute the exact value of the integral. How do the approximations compare to each other and to the exact value of the integral? Do any of the methods compute the exact value of the integral?

1. $\int_0^{10} 2x\,dx$

2. $\int_0^{10} 3x^2\,dx$

3. $\int_0^{10} 4x^3\,dx$

4. $\int_0^{10} 5x^4\,dx$

5. $\int_{0.1}^{2} 6x^{-1}\,dx$

6. $\int_0^{\pi} 7\sin(x)\,dx$

7. $\int_0^{\pi} \sin(8x)\,dx$

8. $\displaystyle\int_1^2 \sqrt{x-1}\,dx$

9. $\displaystyle\int_0^3 e^{-\frac{x}{9}}\,dx$

10. $\displaystyle\int_{-10}^{10} \frac{1}{\sqrt{2\pi}} e^{\frac{-x^2}{2}}\,dx$

7.3 Vector Fields

In Section 7.9 of the text, you were introduced to vector fields. In this section, we'll show you how you can re-use the `quiver` function from Section 2.3 to plot vector fields in MATLAB.

Let's take the following differential equation as our example:

$$x' = 2y$$
$$y' = -2x$$

The following MATLAB inputs use `quiver` to plot the vector field for this differential equation over $-2 \leq x \leq 2$ and $-2 \leq y \leq 2$:

```
>>              [X, Y] = meshgrid(-2:1:2, -2:1:2);

>>              syms x y
>>              dx = vectorize(inline(char(2*y)));
>>              dy = vectorize(inline(char(-2*x)));

>>              arrowX = dx(Y);
>>              arrowY = dy(X);
>>              quiver(X, Y, arrowX, arrowY)
```

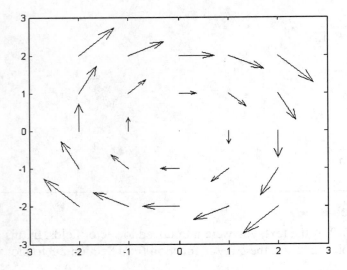

As you can see, you can plot a vector field in almost the same way that you plot a slope field. There are only two minor differences. First, we now have a function for x', so we represent it by the vectorized inline function `dx`. Second, the horizontal component of each arrow is now computed at each point by this `dx` function, just as the vertical component of each arrow is computed at each point by the `dy` function.

There is an important special case, however, when x' or y' are constant functions. For instance, let's suppose that you want to plot the vector field for the following differential equation:

$$x' = y$$
$$y' = 3$$

Since y' is a constant, we need to create the `arrowY` values in the following manner:

```
>>         arrowY = 3 * ones(size(Y));
```

Initially, `ones` sets the y-component of every arrow to 1. Then, we multiply each of these y-components by 3 so that they are equal to y'. In this case, we don't need the `dy` function to create the `arrowY` values.

You're now equipped to plot vector fields!

In the following exercises, use `quiver` to plot the vector field for each of the following differential equations.

1. $x' = 2$
   $y' = -2$

2. $x' = x$
   $y' = -1$

3. $x' = 1 - \sqrt{x}$
   $y' = \sqrt{y}$

4. $x' = \cos(x)$
   $y' = -2$

5. $x' = e^y$
   $y' = y + 1$

6. $x' = .1x^3 + .3xy$
   $y' = .3x - .1x$

7. $x' = e^{.1x - .2y}$
   $y' = x$

7.4 Conclusion

In this chapter, you learned how to approximate integrals using the rectangle, trapezoid, and Simpson methods in MATLAB. You then learned how to plot vector fields in MATLAB using the `quiver` function.

The next chapter will continue along the approximation path and will give you the opportunity to learn some new MATLAB functions.

CHAPTER 8

Polynomial Approximations

In this chapter, we'll begin by showing you how you can use MATLAB to compute Maclaurin and Taylor polynomials for functions. We'll then use a second-order Taylor polynomial to improve the results of the Euler method. Finally, we'll introduce `dsolve`, a MATLAB function that is able to solve some differential equations exactly.

8.1 Maclaurin and Taylor Polynomials in MATLAB

In Chapter 8 of the text, you learned how to compute Maclaurin and Taylor polynomials for functions. In this section, we'll show you how incredibly easy it is to compute these polynomials in MATLAB using the `taylor` function.

Suppose you are trying to find the third-order Taylor polynomial for $g(x) = \frac{1}{x}$ at $x = 2$. Look how easily you can do this in MATLAB.

```
>>          syms x
>>          taylor(1/x, 4, 2)
ans =       1-1/4*x+1/8*(x-2)^2-1/16*(x-2)^3
```

As you can see, `taylor` requires three arguments. The first argument, `1/x`, is the function you want to approximate with a Taylor polynomial. The second argument, `4`, is *one greater* than the order of the Taylor polynomial produced. The last argument, `2`, is the point at which the Taylor polynomial will be computed.

Of course, a Maclaurin polynomial is just a Taylor polynomial computed at $x = 0$, so `taylor` can compute these as well. Here, for instance, is the second-order Maclaurin polynomial for $g(x)=e^x$.

```
>>          taylor(exp(x), 3, 0)
ans =       1+x+1/2*x^2
```

We can see how well a given polynomial approximates a function by plotting the polynomial and the function together. In the next set of inputs, we compare $g(x) = \frac{1}{x}$ to our third-order approximation of it around $x = 2$.

```
>>      taylorPoly = inline(char(taylor(1/x,4,2)), 'x');
>>      ezplot(char(taylorPoly), [0.1, 5])
>>      hold on
>>      ezplot('1/x', [0.1, 5])
>>      hold off
```

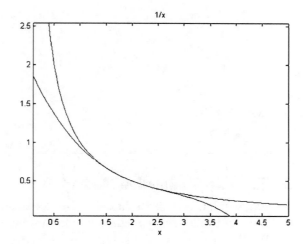

You can see that the Taylor polynomial approximates $g(x) = \frac{1}{x}$ very well around $x = 2$ and poorly elsewhere.

In each of the following exercises, find the first-order, third-order, and fifth-order Maclaurin polynomials for the function $g(x)$. Then make three plots in which $g(x)$ is plotted with each of these approximating Maclaurin polynomials. Be sure to note if any of the Maclaurin polynomials equals $g(x)$.

Hint: A MATLAB script might come in handy for these exercises!

1. $g(x) = (x - 1)^4$
2. $g(x) = (x - 1)^7$

3. $g(x) = \cos(x)$
4. $g(x) = e^{2x}$
5. $g(x) = \frac{1}{x+7}$
6. $g(x) = x^2 \sin(x)$
7. $g(x) = \cos(x) + .2\sin(5x)$

Now, find the first-order, third-order, and fifth-order Taylor polynomials for the function $g(x)$ about the indicated value of x. Then make three plots in which $g(x)$ is plotted with each of these Taylor polynomials. Be sure to note if any of the Maclaurin polynomials equals $g(x)$.

1. $g(x) = x^3$, $x = 1$
2. $g(x) = x^5$, $x = 2$
3. $g(x) = \frac{1}{x}$, $x = -2$
4. $g(x) = \ln(x)$, $x = 2$
5. $g(x) = \sqrt[3]{x}$, $x = 5$
6. $g(x) = \sin(2x) + x$, $x = \pi$
7. $g(x) = \sin(20x) + x$, $x = \pi$

8.2 The Second-Order Euler Approximation Method

In Section 8.9 of the text, you learned about the second-order Euler method. In this section, we'll implement the second-order Euler method by modifying our first-order implementation from Chapter 3. We'll then compare the solutions of the second-order Euler method to those of the first-order Euler method for a number of differential equations.

Our specific goal in this section will be to use the second-order Euler method to compute the value of $f(1)$ from the following information:

$$f' = ft$$
$$f'' = f't + f$$
$$f(0) = 1$$

Let's begin by recalling the basic equation that underlies the second-order Euler method. If we know the value of f at time t_0, then we can approximate the value of f at time $t_0 + \Delta t$ using a second-order Taylor polynomial for f about t_0 as follows:

$$f(t_0 + \Delta t) \approx f(t_0) + f'(t_0)\Delta t + \frac{1}{2}f''(t_0)\Delta t^2$$

Let's also recall our implementation of the first-order Euler method from Chapter 3. Here is the basic script we used, modified to solve the differential equation above. We've also renamed the variable `fEuler1` to `fEuler2`:

```
1               clear fEuler2 dfdt
2
3               tI = 0;
4               tF = 1;
5               dt = 0.25;
6               fInitialValue = 1;
7
8               times = tI:dt:tF;
9
10              fEuler2(1) = fInitialValue;
11              for i = 1:1:(length(times)-1)
12                  dfdt(i) = fEuler2(i)*times(i);
13                  fEuler2(i+1) = fEuler2(i) + dfdt(i)*dt;
14              end
15
16              plot(times, fEuler2)
```

You should be able to see that this script will also work for the second-order Euler method with only minor changes. We just need to compute f'' and add its contribution to `fEuler2` in line 13. We'll leave these minor changes to you as your first exercise for this section.

After making these changes, solve the differential equation above using $\Delta t = 0.25$. You should find that the second-order Euler method approximates the value of $f(1)$ to be 1.6262. You should also find that the approximated curve for $f(t)$ looks like the following graph:

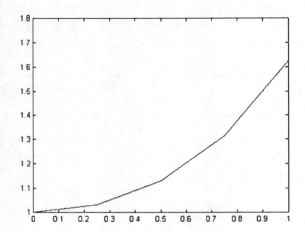

Modify the first-order Euler method implementation above so that it implements the second-order Euler method. Save this new script with the name `euler2Scipt.m`.

Hints:

    a. First, you'll need to add a line between lines 12 and 13.
    b. Next, you'll need to add something to what is now line 13.
    c. Finally, you'll want to add something to line 1.

Now, for each of the following problems, approximate $f(t)$ at the indicated point using both the first-order and second-order Euler method. Compare these values to each other and to the exact value of $f(t)$ at this point. Also, compare the plots of the first-order and second-order Euler method approximations with the plot of the "real" $f(t)$.

1. $\begin{bmatrix} f' = t \\ f(1) = 1 \\ f(2) = ? \end{bmatrix} \Delta t = \dfrac{1}{4}$

2. $\begin{bmatrix} f' = t^2 \\ f(1) = 1 \\ f(2) = ? \end{bmatrix} \Delta t = \dfrac{1}{4}$

3. $\begin{bmatrix} f' = t^3 \\ f(1) = 0.5 \\ f(2) = ? \end{bmatrix} \Delta t = \frac{1}{5}$

4. $\begin{bmatrix} f' = \sin(t) \\ f(0) = 0 \\ f\left(\frac{\pi}{2}\right) = ? \end{bmatrix} \Delta t = \frac{\pi}{8}$

5. $\begin{bmatrix} f = \frac{1}{1+t} \\ f(0) = 1 \\ f(16) = ? \end{bmatrix} \Delta t = 4$

8.3 Solving Differential Equations Exactly with `dsolve`

Until now, we have been finding approximate solutions to differential equations using `ode45`. In this section, we'll show you how to find exact solutions to some differential equations using MATLAB's `dsolve` function.

Suppose you are trying to solve the following system of differential equations with the given initial conditions:

$$x' = y$$
$$y' = x$$

$$x(0) = 2$$
$$y(0) = 1$$

You can find the exact solutions for $x(t)$ and $y(t)$ using MATLAB's `dsolve` function as follows:

```
>>              [x, y] = dsolve('Dx = y', 'Dy = x', 'x(0) = 2', 'y(0) = 1')
x =
                1/2*exp(-t)+3/2*exp(t)
y =
                3/2*exp(t)-1/2*exp(-t)
```

Presto! The function `dsolve` produces the exact solutions for $x(t)$ and $y(t)$.

As you can see, there is one argument to `dsolve` for each differential equation and initial value condition. Each of these equations is contained within single quotations and is separated from the next equation by a comma. Differentials are expressed with the D character. Thus, x' is written as Dx. Initial value conditions are expressed in a straightforward way.

To evaluate and plot dsolve's solutions for x(t) and y(t), you need to explicitly turn dsolve's outputs into inline functions.

```
>>              xFunc = inline(x);
>>              yFunc = inline(y);
```

Now, for instance, you can verify that these solutions for x(t) and y(t) satisfy the initial value conditions.

```
>>              xFunc(0)
ans =           2
>>              yFunc(0)
ans =           1
```

You can also plot the solutions for x(t) and y(t) in the usual way, as follows:

```
>>              syms t
>>              ezplot(char(xFunc(t)), [-4, 2, -7, 12])
>>              hold on, grid
>>              ezplot(char(yFunc(t)), [-4, 2, -7, 12])
>>              title('x(t) and y(t)')
>>              hold off
```

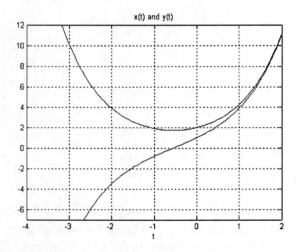

As impressive as dsolve is, a word of caution is in order. In the grand scheme of differential equations, the number of differential equations that dsolve can actually solve is quite miniscule. If you give dsolve a complicated set of differential equations, you may find that dsolve generates a barrage of warning and error messages and gives up. This is exactly what happened when we tried to use dsolve to solve an "Epidemic at the Track" in Section 4.4, for example.

When things like this happen, this should be your cue to solve the differential equations approximately with ode45.

Solve the following sets of differential equations exactly using dsolve. Plot the solutions for $0 \leq t \leq 1$ and compute the values of $x(1)$ and $y(1)$.

1. $\begin{bmatrix} x' = y \\ y' = x \end{bmatrix}$
   $x(0) = 1$
   $y(0) = 1$

2. $\begin{bmatrix} x' = y \\ y' = x \end{bmatrix}$
   $x(0) = 1$
   $y(0) = -1$

3. $\begin{bmatrix} x' = y \\ y' = -x \end{bmatrix}$
   $x(0) = 1$
   $y(0) = 1$

4. $\begin{bmatrix} x' = y \\ y' = -x \end{bmatrix}$
   $x(0) = 1$
   $y(0) = -1$

5. $\begin{bmatrix} x' = y \\ y' = -2x \end{bmatrix}$
   $x(0) = 2$
   $y(0) = 1$

6. $\begin{bmatrix} x' = \frac{1}{3}y \\ y' = x \\ x(0) = 1 \\ y(0) = 0 \end{bmatrix}$

Now, solve the following differential equations using both the first-order and second-order Euler method. Then solve the differential equations exactly using `dsolve`. Plot the three solutions and compare their results.

Note: In each exercise, use $\Delta t = \frac{3}{5}$.

1. $\begin{bmatrix} f' = t + f \\ f(0) = 1 \\ f(3) = ? \end{bmatrix}$

2. $\begin{bmatrix} f' = 2t + f \\ f(0) = 1 \\ f(3) = ? \end{bmatrix}$

3. $\begin{bmatrix} f' = t - f \\ f(0) = 1 \\ f(3) = ? \end{bmatrix}$

4. $\begin{bmatrix} f' = t - 2f \\ f(0) = 1 \\ f(3) = ? \end{bmatrix}$

5. $\begin{bmatrix} f' = t^2 - f \\ f(0) = 1 \\ f(3) = ? \end{bmatrix}$

6. $\begin{bmatrix} f' = tf \\ f(0) = 1 \\ f(3) = ? \end{bmatrix}$

7. $\begin{bmatrix} f' = t^2 f \\ f(0) = 1 \\ f(3) = ? \end{bmatrix}$

8. $\begin{bmatrix} f' &=& \dfrac{t}{f} \\ f(0) &=& 1 \\ f(3) &=& ? \end{bmatrix}$

9. $\begin{bmatrix} f' &=& f^2 \\ f(0) &=& 1 \\ f(3) &=& ? \end{bmatrix}$

10. $\begin{bmatrix} f' &=& \sqrt{f} \\ f(0) &=& 1 \\ f(3) &=& ? \end{bmatrix}$

8.4 Conclusion

In this chapter, you learned how to use `taylor` to create Maclaurin and Taylor polynomials that approximate functions. You then used a second-order Taylor polynomial to improve the solutions computed by the Euler Method for differential equations. Finally, you learned how to find exact solutions to some differential equations using MATLAB's `dsolve` function.

In the next chapter, you'll move from the realms of the approximate to the realms of the exact as you continue your study of Maclaurin and Taylor polynomials.

CHAPTER 9

Infinite Series

In Chapter 9 of the textbook, you learned about infinite series. In this chapter, we cover the relevant MATLAB functions and techniques that you'll need to study the infinite in MATLAB. We also give you plenty of opportunities to sharpen your skills with respect to infinite series, radii and intervals of convergence, L'Hôpital's rule, and improper integrals.

9.1 `symsum` and Series

To study series in MATLAB, you'll need to be comfortable with the function `symsum`. You've already run across `symsum` in Chapter 5, but we'll review how it works in this section.

Suppose you want to compute the following summation:

$$\sum_{k=1}^{k=20} \frac{1}{k^5}$$

As much fun as it would be to compute this by hand, you can find the answer much more quickly by plugging this summation into `symsum`:

```
>>        syms k
>>        symsum(1/k^5, k, 1, 20)
ans =     18904366180115796232527531782784402865\
          21 / 182311562617504492965760613453054607\
          3600
```

Never one to underachieve, MATLAB even computes the *exact* value of the summation. Of course, if you don't mind trading exactness for succinctness, you can filter the output of `symsum` through the `double` function.

```
>>        double(symsum(1/k^5, k, 1, 20))
ans =     1.0369
```

Now suppose we are curious about whether the following series converges and, if so, to what value it converges.

$$a_n = \left(-\frac{5}{6}\right)^n$$

We can get a sense for both answers by plotting the series for $n = 1$ to $n = 20$.

```
>>   syms k n
>>   a = vectorize(inline(char(symsum((-5/6)^k, k, 1, n))));
>>   N = 1:1:20;
>>   A = a(N);
>>   plot(N, A, 'x')
>>   xlabel('n'), ylabel('Sum of a(n)')
```

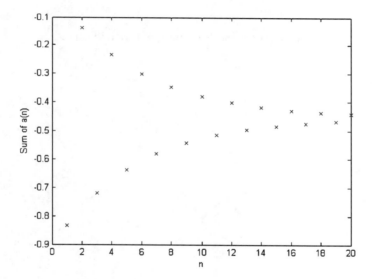

From this plot, it's pretty clear that the series *is* converging, and it appears to be converging to a value between -0.4 and -0.5. In fact, MATLAB can tell us the *exact* value to which this series converges if we ask what the series' value is when n goes to infinity.

```
>>   symsum((-5/6)^n, n, 1, inf)
ans =
     -5/11
```

Impressive, no?

Plot the first twenty or more terms of the following series and decide whether each series is converging. If you think the series is converging, estimate to which value it is converging. Finally, confirm your estimate by using MATLAB to compute the exact value to which the series converges.

Note: In each series, the first value of n is 1.

1. $a_n = 0$
2. $a_n = 1$
3. $a_n = \dfrac{1}{google}$ (Recall that google = 10^{100})
4. $a_n = \dfrac{(-1)^n}{n}$
5. $a_n = \dfrac{(-1)^n}{n^2}$

Now, carry out the same steps for the following geometric series.

1. $a_n = 0^n$
2. $a_n = 1^n$
3. $a_n = (-1)^n$
4. $a_n = \left(\dfrac{9}{10}\right)^n$
5. $a_n = \left(\dfrac{10}{9}\right)^n$
6. $a_n = \left(\dfrac{999999}{1000000}\right)^n$
7. $a_n = \left(\dfrac{1000000}{999999}\right)^n$
8. $a_n = \left(-\dfrac{9}{10}\right)^n$
9. $a_n = \left(-\dfrac{10}{9}\right)^n$
10. $a_n = \dfrac{9}{10^n}$

What condition must be met in order for a geometric series to converge?

9.2 Intervals and Radii of Convergence

In this section, you'll use MATLAB to visualize intervals and radii of convergence for series. You've already learned how to compute and plot Maclaurin and Taylor series in Chapter 8, so we'll just show you a convenient way to plot general polynomial series.

Infinite Series

Suppose you want to plot several approximations of the following polynomial series:

$$a_k = \sum_{k=1}^{\infty} (5x + 2)^k$$

We can approximate this series by the polynomial formed by its first n terms. The following inputs define a function that creates this very polynomial for any value of n:

```
>>              syms x n k
>>              a = inline(char(symsum((5*x+2)^k, k, 1, n)), 'x', 'n');
```

Now, for example, we can plot the polynomial approximation of this series that results from summing its first three terms:

```
>>              ezplot(char(a(x,3)), [-1, 1, -10, 10])
>>              grid, title('a(x,3)');
```

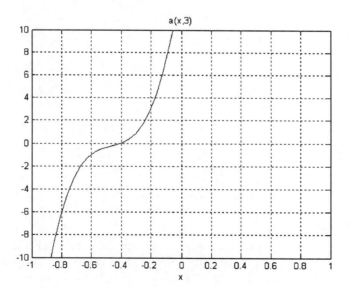

This technique will be useful to you in the second set of exercises in this section.

In the following exercises, use MATLAB to compute the Maclaurin or Taylor series for each function. Then, using paper and pencil, determine the interval and the radius of convergence of each series. Finally, confirm this interval and radius by plotting the function with the third-, seventh-, and eleventh-order Maclaurin or Taylor series for the function.

1. The Maclaurin series for $f(x) = e^x$
2. The Maclaurin series for $f(x) = xe^x$
3. The Taylor series about $x = 3$ for $f(x) = \ln(x)$
4. The Maclaurin series for $f(x) = \frac{1}{1-x}$
5. The Maclaurin series for $f(x) = \frac{x^2}{1-x}$

In the next set of exercises, determine the interval and the radius of convergence of each series. Then, confirm this interval and radius by plotting the third-, seventh-, and eleventh-order approximations of the series.

1. $\sum_{n=0}^{\infty} (2x - 3)^n$
2. $\sum_{n=0}^{\infty} \frac{1}{4}(3x - 1)^n$
3. $\sum_{n=0}^{\infty} \frac{1}{n!} x^n$
4. $\sum_{n=3}^{\infty} \frac{1}{n!} x^n$
5. $\sum_{n=0}^{\infty} (-1)^n \frac{x^{2n}}{(2n)!} + \sum_{n=0}^{\infty} (-1)^n \frac{x^{2n+1}}{(2n + 1)!}$

9.3 L'Hôpital's Rule

In Section 9.8 of the textbook, you were introduced to L'Hôpital's rule. In this section, you'll use L'Hôpital's rule and MATLAB to compute and verify the limits of indeterminate forms. Since you already know all the MATLAB you'll need to do these exercises, it's time to let you loose on them!

Find the following limits using L'Hôpital's rule whenever it applies. Then plot each function to confirm that the limit you found is reasonable.

1. $\lim\limits_{x \to 1} \dfrac{x^2 - 1}{x - 1}$

2. $\lim\limits_{x \to 0} \dfrac{-3x^2 - 2x}{x^2 + 4x}$

3. $\lim\limits_{x \to \infty} \dfrac{-3x^2 - 2x}{x^2 + 4x}$

4. $\lim\limits_{x \to 0} \dfrac{-3x^2 - 2x}{x^4 + 4x}$

5. $\lim\limits_{x \to \infty} \dfrac{-3x^2 - 2x}{x^4 + 4x}$

6. $\lim\limits_{x \to 0} \dfrac{\sin(x) + x}{x}$

7. $\lim\limits_{x \to 0} \dfrac{\cos\left(x + \dfrac{\pi}{2}\right) - x}{x}$

8. $\lim\limits_{x \to 0} \dfrac{x\cos(x)}{\sin(x)}$

9. $\lim\limits_{x \to \infty} \dfrac{\ln(x)}{e^x}$

10. $\lim\limits_{x \to \infty} \dfrac{\ln(x)}{\sqrt{x}}$

9.4 Improper Integrals

MATLAB is not intimidated when it sees infinity in an integral, whether the infinity is in the limits of the integral or in the integrand itself. For example, in the next two inputs, we evaluate the following integral:

$$\int_1^\infty \frac{1}{x^5}\, dx$$

```
>>                      syms x
>>                      int(1/x^5, x, 1, inf)
ans =                   1/4
```

However, many such integrals can't be evaluated. Take the following integral, for example:

$$\int_1^\infty \frac{1}{x}\,dx$$

This integral does not have a value, and MATLAB tells us this when we try to integrate it.

```
>>                      int(1/x, x, 1, inf)
ans =                   inf
```

In the next set of exercises, you'll use MATLAB to gain some intuition about improper integrals.

In each of the following exercises, an improper integral is being computed. First, plot each integrand over a large enough domain to see what it looks like over the limits of integration. Then, compute the value of the integral using `int`.

Note: Filter the results of `int` through MATLAB's `double` function to compute the decimal value of each of these integrals.

1. $\displaystyle\int_{100}^\infty \frac{1}{x^1}\,dx$

2. $\displaystyle\int_1^\infty \frac{1}{x^2}\,dx$

3. $\displaystyle\int_1^\infty \frac{1}{x^{1.01}}\,dx$

4. $\displaystyle\int_1^\infty \frac{4}{(1+x^2)}\,dx$

5. $\displaystyle\int_1^\infty \frac{4x^1}{(1+x^2)}\,dx$

6. $\displaystyle\int_1^\infty \frac{4x^{0.6}}{(1+x^2)}\,dx$

7. $\displaystyle\int_0^\infty x^0 e^{-x}\,dx$

8. $\displaystyle\int_0^\infty x^1 e^{-x}\,dx$

9. $\displaystyle\int_0^\infty x^2 e^{-x}\,dx$

10. $\displaystyle\int_{-\infty}^\infty x^0 e^{-x^2}\,dx$

11. $\displaystyle\int_0^\infty x^1 e^{-x^2}\,dx$

12. $\displaystyle\int_0^\infty x^2 e^{-x^3}\,dx$

13. $\displaystyle\int_0^\infty \cos(x)\,dx$ (Note how MATLAB handles this integral.)

14. $\displaystyle\int_0^\infty e^{-x}\cos(x)\,dx$

15. $\displaystyle\int_1^2 \frac{1}{(x-1)^1}\,dx$

16. $\displaystyle\int_{1}^{2} \frac{1}{(x-1)^{0.5}}\,dx$

17. $\displaystyle\int_{1}^{2} \frac{1}{(x-1)^{0.99}}\,dx$

9.5 Conclusion

In this chapter, we confronted the infinite with MATLAB. We saw that MATLAB is adept at working with infinite series and computing improper integrals that involve infinity. We also saw that MATLAB is useful for visualizing the convergence of infinite series and confirming the limits computed by L'Hôpital's rule.

Visualizing problems is one of the great advantages of using MATLAB, and in the next chapter we'll learn how to use MATLAB to help us visualize functions in three dimensions.

CHAPTER 10

The Third Dimension

In Chapter 10 of the textbook, you learned about two-variable functions. In particular, you learned how to plot them, differentiate them, and integrate them. You probably won't be surprised to hear that MATLAB can do all these things too, and our goal in this chapter will be to show you how.

10.1 Picturing Functions of Two Variables

MATLAB can plot functions of two variables much more quickly than a human can. In this section, we'll show you how to use MATLAB to draw surfaces and contour maps for two-variable functions.

Throughout this section, we'll use the following two-variable function as our example:

$$z = F(x,y) = x^2 - y^2$$

10.1.1 Defining a Two-Variable Function

First and foremost, let's define the two-variable function $F(x,y)$ in MATLAB.

```
>>              syms x y
>>              F = vectorize(inline(char(x^2-y^2), 'x', 'y'));
```

Generally speaking, it's a good idea to define two-variable functions with `vectorize` since some of MATLAB's plotting functions require it, as we'll see.

You can now evaluate F in an intuitive way. For example, here's the value of F when $x = 3$ and $y = 2$:

```
>>              F(3, 2)
ans =           5
```

10.1.2 Plotting Surfaces

We'll now show you how to use MATLAB's `ezsurf` and `ezmesh` functions to draw a three-dimensional surface for $F(x,y)$.

We'll begin with the `ezsurf` function, which works just like `ezplot`. The only difference is that `ezsurf` expects its first argument to be a two-variable function.

```
>>      ezsurf(char(F(x,y)), [-2, 2, -3, 3])
>>      colorbar
```

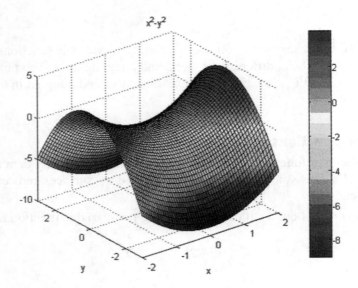

As with `ezplot`, the first argument to `ezsurf` is a string representation of the function to be plotted. The second argument, [−2, 2, −3, 3], specifies the plot domain for the independent variables. In this case, x ranges between −2 and 2, and y ranges between −3 and 3.

The surface drawn by `ezsurf` is colored according to the value of $F(x,y)$. The larger values of $F(x,y)$ are colored with warmer colors, and the smaller values are colored with cooler colors. You can display the quantitative mapping of color to the value of $F(x,y)$ using MATLAB's `colorbar` function.

To control the direction from which you view the surfaces drawn by `ezsurf`, you can use MATLAB's `view` function. For example, when you execute the following input, you'll see the surface for $F(x,y)$ from a different direction.

```
>>      view([1, 0.25, 0.15])
```

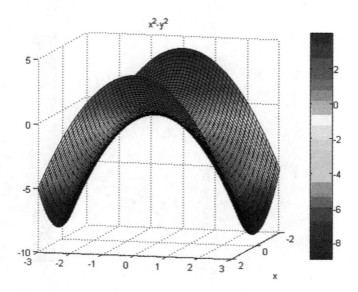

The argument to view specifies the *x, y,* and *z* components of a vector, and this vector determines the direction from which you view the surface. In this call to view, the *x* component of the vector is 1, the *y* component is 0.25, and the *z* component is 0.15. Thus, MATLAB redraws the surface for $F(x,y)$ so that we are looking at it from slightly off the *x*-axis.

Finally, you can also create a mesh representation of $F(x,y)$ using MATLAB's ezmesh function. The arguments of ezmesh are exactly the same as the arguments of ezplot.

```
>>          ezmesh(char(F(x,y)), [-2, 2, -3, 3])
>>          colorbar
```

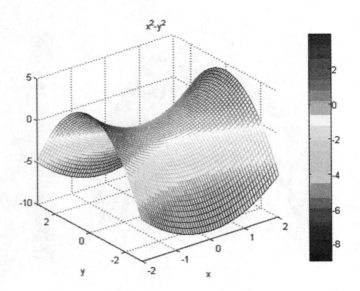

Quite frankly, the choice between `ezsurf` and `ezmesh` is strictly a matter of aesthetics. There's nothing you can learn from an `ezmesh` plot that you can't learn from an `ezsurf` plot!

Before we move on, we'll briefly note that there are other MATLAB functions for plotting surfaces. For instance, just as `ezplot` has a corresponding `plot` function, `ezsurf` has a corresponding `surf` function. However, `ezsurf` can do everything we'll need a surface plotter to do in this chapter, and it *is* very easy to use. Thus, `ezsurf` will be our surface plotting function of choice in MATLAB.

Unless, of course, you think `ezmesh`'s plots are prettier.

10.1.3 Plotting Contour Maps

Contour maps are another way to picture functions of two variables, and MATLAB draws them using the functions `ezcontourf` and `contourf`. In the case of `ezcontourf` and `contourf`, there *are* good reasons to learn how to use both, so we'll cover both functions in this section.

In the next two inputs, we use `ezcontourf` to draw the contour map for our function $F(x,y)$.

```
>>      ezcontourf(char(F(x,y)), [-2, 2, -3, 3])
>>      colorbar
```

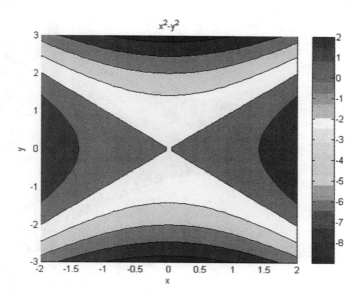

As you can see, the arguments of `ezcontourf` are exactly the same as the arguments of `ezsurf` and `ezmesh`. The color scheme is also the same — higher regions of F are colored with warmer colors, while lower regions of F are colored with cooler colors. You can again use the `colorbar` function to display the quantitative mapping of color to the value of $F(x,y)$.

The "problem" with `ezcontourf` is that it automatically decides which contour lines to draw, and there's no way to change this decision. On the other hand, MATLAB's `contourf` function lets you decide exactly which contour lines are drawn. This is a useful enough feature that we'll show you how to use the `contourf` function, even if it isn't as easy to use as `ezcontourf`.

The following inputs use `contourf` to plot the contour map for $F(x,y)$ with contour lines where $F(x,y)$ equals 1, 0, and -1.

```
>>          [X, Y] = meshgrid(-2:.1:2, -3:.1:3);
>>          contourf(X, Y, F(X,Y), [-1, 0, 1])
>>          colorbar, xlabel('x'), ylabel('y')
```

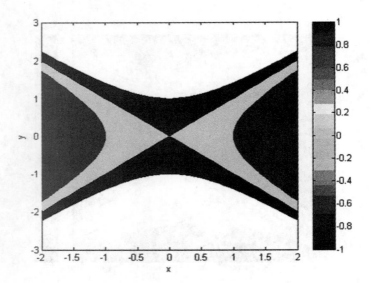

In the first input, we create the two-dimensional grid of points over which the contour map will be plotted. In this case, we're plotting the contour map for F over the rectangular grid bounded by $-2 \leq x \leq 2$ and $-3 \leq y \leq 3$.

In the second input, we use `contourf` to actually draw the contour lines for F over this two-dimensional grid. The first two arguments, X and Y, define the x- and y-coordinates of the two-dimensional grid. The third argument, F(X,Y), computes the values of F at these points using the vectorized F function. The last argument, [-1,0,1], tells `contourf` to draw contour lines where F equals 1, 0, or -1.

You can also set the fourth argument of `contourf` to a number, and `contourf` will use this number of contour lines in its contour map for F. For example, in the next input we tell `contourf` to draw five contour lines for F.

```
>>          contourf(X, Y, F(X,Y), 5)
>>          colorbar, xlabel('x'), ylabel('y')
```

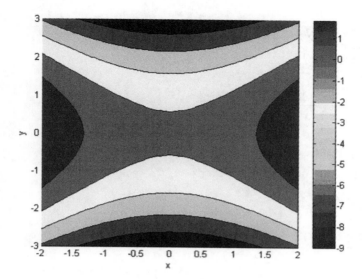

Note that when you use `contourf` you have to manually label the axes with `xlabel` and `ylabel`. However, `ezcontourf` does this automatically.

For each of the following functions of two variables, use `ezsurf` or `ezmesh` to plot the corresponding surface and `ezcontourf` or `contourf` to plot the corresponding contour map. Try to make the most visually informative (and aesthetically pleasing) pictures possible.

1. $F(x,y) = 2$
2. $F(x,y) = 2x$
3. $F(x,y) = 2x - y$
4. $F(x,y) = 2x^2$
5. $F(x,y) = 2x^2 + y^2$
6. $F(x,y) = 2x^2 + y^3$
7. $F(x,y) = \dfrac{x^3}{1 + y^2}$
8. $F(x,y) = 2\sin(x)$
9. $F(x,y) = 2\sin(x) - \cos(y)$
10. $F(x,y) = 2\sin(x)\cos(y)$
11. $F(x,y) = y\sin(x)$

12. $F(x,y) = \dfrac{1}{1+y^2}$

13. $F(x,y) = \dfrac{1}{1+x^2+y^2}$

14. $F(x,y) = \dfrac{1}{2+|x|+|y|}$

15. $F(x,y) = e^{-x}$

16. $F(x,y) = ye^{-x}$

17. $F(x,y) = e^{-(x+y)}$

18. $F(x,y) = e^{-(x^2+y^2)}$

19. $F(x,y) = e^{-\frac{x^2y^2}{20}}$

20. $F(x,y) = \cos(x)\sin(2y)e^{-\frac{x^2y^2}{20}}$

10.2 Differentiating Functions of Two Variables

In this section, we'll show you how to use MATLAB's `diff` function to take partial derivatives of functions of two variables. We'll also show you several useful ways to manipulate partial derivatives after you've computed them.

Let's take the following function as our example:

$$F(x,y) = (y+1)^2(x-1)$$

For convenience, we begin by defining this function in MATLAB.

```
>>          syms x y
>>          F = vectorize(inline(char((y+1)^2*(x-1)), 'x', 'y'));
```

Now we can use MATLAB's `diff` function to compute the partial derivative of F with respect to x.

```
>>          diff(F(x,y), x)
ans =       (y+1)^2
```

We can also compute the partial derivative of F with respect to y in a similar way.

```
>>          diff(F(x,y), y)
ans =       2*(y+1)*(x-1)
```

Now, suppose we want to evaluate these partial derivatives at particular values of x and y. The easiest way to do this is to first create inline functions for the partial derivatives.

```
>>      dFdx = vectorize(inline(char(diff(F(x,y), x)), 'x', 'y'));
>>      dFdy = vectorize(inline(char(diff(F(x,y), y)), 'x', 'y'));
```

Note that even though the partial derivative of F with respect to x is actually independent of x, it's OK to make x an argument of dFdx. MATLAB won't complain, and dFdx will still compute the correct values for the partial derivative.

Now, we can use dFdx and dFdy to evaluate the partial derivatives at any point we like.

```
>>      dFdx(1,1)
ans =   4
>>      dFdy(0,1)
ans =   -4
```

Finally, suppose you want to find the points where a partial derivative is equal to zero. MATLAB's solve function provides a quick and easy way to do this. For example, to find the points where dFdy[x,y] equals zero, you can use the following input:

```
>>      [xSoln,ySoln] = solve('2*(y+1)*(x-1) = 0', 'x,y')
xSoln = [x]
        [1]
ySoln = [-1]
        [y]
```

As you can easily verify, dFdy[x,y] does indeed equal zero when $x = 1$ or $y = -1$.

Another way to find the zeros of a partial derivative is to make a contour map of it. For instance, in the next input we make a contour map of dFdy with contour lines drawn where dFdy equals -1, 0, and 1.

```
>>      [X, Y] = meshgrid(-2:.1:2, -2:.1:2);
>>      contourf(X, Y, dFdy(X,Y), [-1, 0, 1])
>>      colorbar, xlabel('x'), ylabel('y')
```

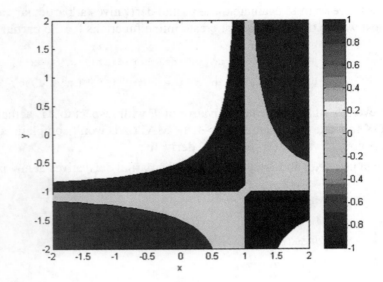

This contour map makes it pretty clear that dFdy has zeros along the lines $x = 1$ and $y = -1$.

In the next set of exercises, we'll use what you learned in this section to sharpen your partial derivative skills.

First, plot each $F(x,y)$ as a surface and a contour map in the neighborhood of the given points. Then, based on your plots, decide if the partial derivative of F_x is positive, negative, or zero at each point. Finally, confirm your decisions by using MATLAB to compute the actual value of F_x at each point. Use a contour plot of F_x to confirm your decisions as well.

1. $F(x,y) = 2x^2 + y^3$ at $(0,-2)$ and $(1,1)$
2. $F(x,y) = \dfrac{x^3}{1 + y^2}$ at $(1,-1)$ and $(0,2)$
3. $F(x,y) = y\sin(x)$ at $(1,0)$ and $(1,-2)$
4. $F(x,y) = 2\sin(x)\cos(y)$ at $(0,1)$ and $(1,1)$
5. $F(x,y) = \sin(xy)$ at $(1,-1)$ and $(1,1)$ and $\left(1, \dfrac{\pi}{2}\right)$
6. $F(x,y) = e^{-(x^2+y^2)}$ at $(1,1)$ and $(0,0)$

Use MATLAB's `diff` and `solve` functions to find the critical points for each of the following functions. Then plot surface and contour maps for each of the functions to confirm that you have identified all the critical points. Once you are sure that you've found all the critical points, decide whether they are local maxima, minima, or saddle points. You may also find it useful to make contour maps of the partial derivative functions to locate critical points.

1. $F(x,y) = 8 - 6x + x^2 - 2y - y^2$
2. $F(x,y) = 9 - 6x + x^2 + 18y - 12xy + 2x^2y + 9y^2 - 6xy^2 + x^2y^2$
3. $F(x,y) = y^2 - 2xy^2 + x^2y^2$
4. $F(x,y) = \dfrac{x}{x^2 + y^2 + 1}$
5. $F(x,y) = \cos(y)$
6. $F(x,y) = y\sin(2x)$
7. $F(x,y) = \sin(xy)$
8. $F(x,y) = e^{-(x+y)}$
9. $F(x,y) = \cos(x)\sin(y)e^{-x}$
10. $F(x,y) = \dfrac{1}{2}\sqrt{x^{10} + y^{10}} - \sqrt{x^8 + y^8}$

10.3 Integrating Functions of Two Variables

In MATLAB, integrating functions of two variables is almost as easy as differentiating functions of two variables. In the same way that we could re-use `diff` to differentiate functions of two variables, we can re-use `int` to integrate functions of two variables.

We'll use the following function as our example:

$$F(x,y) = x^2 - xy + 3$$

Let's suppose that we want to evaluate the following definite integral of this function:

$$\int_0^3 \int_{-1}^2 F(x,y)\,dx\,dy$$

We can compute this integral using the `int` function twice, as follows:

```
>>          syms x y
>>          F = vectorize(inline(char(x^2-x*y+3), 'x', 'y'));
>>          int( int(F(x,y), x, -1, 2), y, 0, 3)
ans =       117/4
```

The innermost call to `int` evaluates the integral with respect to x. Then, the outermost call to `int` integrates this result with respect to y.

You can also compute the value of definite integrals that have variables in their limits of integration. For instance, to compute the value of the integral

$$\int_0^3 \int_{x_1}^{x_2} F(x,y)\,dx\,dy$$

you can use the following inputs:

```
>>          syms x1 x2
>>          int( int(F(x,y), x, x1, x2), y, 0, 3)
ans =       -9/4*x2^2+9/4*x1^2+x2^3-x1^3+9*x2-9*x1
```

Finally, you can also evaluate indefinite double integrals in the same way. For example, to evaluate the following indefinite integral:

$$\iint F(x,y)\,dx\,dy$$

you can execute the following input:

```
>>          int( int(F(x,y), x), y)
ans =       1/3*x^3*y-1/4*x^2*y^2+3*x*y
```

Plot each of the following functions over the given domains for x and y. Then, compute the volume below each of the functions over these domains.

1. $H(x,y) = 1$ $-2 \le x \le 2, 5 \le y \le 10$
2. $H(x,y) = y$ $-3 \le x \le 0, 7 \le y \le 9$
3. $H(x,y) = x + y$ $1 \le x \le 5, 0 \le y \le 4$
4. $H(x,y) = x^2 + y^2$ $1 \le x \le 3, 0 \le y \le 2$
5. $H(x,y) = \cos(xy)$ $0 \le x \le \sqrt{\pi}, 0 \le y \le \sqrt{\pi}$

6. $H(x,y) = \dfrac{x}{x^2 + y^2 + 1}$ $\qquad 0 \le x \le 4,\ -2 \le y \le 2$

7. $H(x,y) = \dfrac{1}{2\pi} e^{-\frac{1}{2}(x^2+y^2)}$ $\qquad -2 \le x \le 2,\ -2 \le y \le 2$

and

$-5 \le x \le 5,\ -5 \le y \le 5$

10.4 Conclusion

In this chapter we have introduced the MATLAB functions you need to be able to work with functions of two variables. We began by showing you how to plot surfaces and contour maps of these functions using `ezsurf`, `ezcontourf`, and `contourf`. We then showed you how easily functions of two variables can be differentiated and integrated in MATLAB using `diff` and `int`.

Believe it or not, we will introduce still more ways to plot functions in MATLAB in the next chapter.

CHAPTER 11
Polar, Cylindrical, and Spherical Coordinates

In Chapter 11 of the textbook, you were introduced to polar, cylindrical, and spherical coordinates. In this chapter, we'll show you how to use these different coordinate systems in MATLAB.

11.1 Using Polar Coordinates in MATLAB

In this section, we'll show you how to use MATLAB to plot functions expressed in polar coordinates. We'll also show you how you can use MATLAB to convert between polar coordinates and rectangular coordinates.

11.1.1 Plotting in Polar Coordinates

In this section, we'll show you how easy it is to plot polar functions in MATLAB using MATLAB's `ezpolar` function.

Let's take the following polar function as our example:

$$r = \sin(\theta^2)$$

For convenience, we'll start by defining this function in MATLAB.

```
>>          syms theta
>>          r = vectorize(inline(char(sin(theta^2))));
```

Now, we can use `ezpolar` to plot $r = \sin(\theta^2)$.

```
>>          ezpolar(char(r(theta)), [0, pi])
```

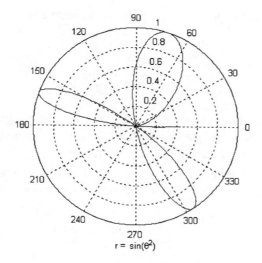

$r = \sin(\theta^2)$

As you can see, `ezpolar` works just like the rest of the `ez` plotting family, although the resulting plot is tailored specifically to polar coordinates.

Believe it or not, that's all there is to plotting functions in polar coordinates!

11.1.2 Converting Between Rectangular Coordinates And Polar Coordinates

You may find, if only in a problem set, that you need to convert between polar coordinates and rectangular coordinates. In this section, we'll show you how MATLAB's `solve` function can do this conversion for you.

Let's suppose we want to convert the following function to polar coordinates:

$$y = 2x - 3$$

We can do this conversion in MATLAB by first defining y and x in terms of their polar equivalents.

```
>>          syms x y r theta
>>          x = r * cos(theta);
>>          y = r * sin(theta);
```

Now we ask `solve` to solve $y = 2x - 3$ for r as follows:

```
>>          solve(y - 2*x + 3, 'r')
ans =       3/(-sin(theta)+2*cos(theta))
```

Voilà! The `solve` function produces the polar coordinate equivalent of the function $y = 2x - 3$:

$$r = \frac{3}{-\sin(\theta) + 2\cos(\theta)}$$

Note that you could also have used `2*x - 3 - y` as the first argument of `solve`. This argument just needs to be the rectangular coordinate function, expressed with all of the variables and constants collected on one side of the equation.

You can also convert polar coordinates to rectangular coordinates in an analogous manner. For example, in the next set of inputs we use `solve` to convert the polar function

$$r = \frac{3}{-\sin(\theta) + 2\cos(\theta)}$$

back to rectangular coordinates.

```
>>     syms x y r theta
>>     r = sqrt(x^2 + y^2);
>>     theta = atan(y/x);
>>     solve(r - 3/(-sin(theta)+2*cos(theta)), 'y')
ans =
       [i*x]
       [-i*x]
       [2*x+3]
       [2*x-3]
```

While `solve` does find the solution we expect, $y = 2x - 3$, it also finds three other candidate solutions. We'll leave it to you to decide if these other solutions are actually valid!

Use `ezpolar` to plot the following functions over the given domain for θ.

1. $r = 2$ $0 \leq \theta \leq \pi$
2. $r = \theta$ $0 \leq \theta \leq 3\pi$
3. $r = -\theta$ $0 \leq \theta \leq 3\pi$
4. $r = \cos(\theta)$ $0 \leq \theta \leq \pi$
5. $r = \sin(\theta)$ $0 \leq \theta \leq \pi$
6. $r = \cos(30\theta)$ $0 \leq \theta \leq \pi$

7. $r = \cos(\theta^2)$ $0 \le \theta \le \pi$
8. $r = \cos(\theta)\sin(\theta)$ $0 \le \theta \le 2\pi$
9. $r = \sec(\theta) + 2\cos(\theta)$ $0 \le \theta \le \frac{29}{60}\pi$
10. $r = \sec(\theta) + 2\cos(\theta)$ $0 \le \theta \le \frac{30}{60}\pi$

Use `ezpolar` to plot the following functions over the given domain for θ. Then use `int` to compute the area of the region enclosed by the arc and the bounds on θ.

1. $r = \frac{1}{2\pi}$ $0 \le \theta \le 2\pi$
2. $r = \theta$ $0 \le \theta \le \pi$
3. $r = \cos(3\theta)$ $0 \le \theta \le \pi$
4. $r = 1 - \sin(\theta)$ $0 \le \theta \le 2\pi$
5. $r = \sin(\theta^2)$ $0 \le \theta \le \sqrt{\pi}$

Convert the following functions in rectangular coordinates into functions in polar coordinates. Plot both versions of the functions to confirm that they are one and the same. Note how much more compactly certain functions can be written in rectangular coordinates compared to polar coordinates.

1. $y = 1$
2. $y = 1 + x$
3. $y = x^2$
4. $y = x^3$
5. $y = \frac{1}{x}$
6. $y = \sqrt{x}$
7. $y = \sqrt{1 + x^2}$
8. $x^2 + y^2 = 1$
9. $(x - 1)^2 + y^2 = 1$
10. $(x - 1)^2 + y^2 = 9$

Convert the following functions in polar coordinates into functions in rectangular coordinates. Plot the polar versions and do your best to plot the rectangular coordinate versions, although in some cases this will be quite challenging. At the very least, appreciate the conciseness of the polar coordinate representation!

1. $r = 1$
2. $r = \sin(\theta)$
3. $r = \cos(\theta)$
4. $r = 1 - \cos(\theta)$
5. $r = 1 - \sin(\theta)$

11.2 Plotting in Cylindrical and Spherical Coordinates

Plotting functions in cylindrical and spherical coordinates isn't straightforward because MATLAB's surface plotting functions can only plot in rectangular coordinates. Thus, in this section, we'll show you how to convert a function expressed in cylindrical or spherical coordinates to rectangular coordinates and then plot its surface.

11.2.1 Plotting in Cylindrical Coordinates

Suppose we want to plot the following cylindrical coordinate function in MATLAB:

$$z = r\sin^2(\theta), \quad 0 \leq \theta \leq 1, \quad 0 \leq \theta \leq 2\pi$$

Since MATLAB's surface plotting functions require rectangular coordinates, we must use the following three steps to plot the surface for this function in MATLAB:

1. Compute the points on the surface of $z = r\sin^2(\theta)$ in cylindrical coordinates.
2. Convert these points from cylindrical coordinates to rectangular coordinates.
3. Use MATLAB to plot the surface defined by these rectangular coordinates.

Step 1

To carry out the first step, we begin by defining the function $z = r\sin^2(\theta)$ in MATLAB.

```
>>    syms r theta
>>    z = vectorize(inline(char(r*sin(theta)^2), 'r', 'theta'));
```

Next, we evaluate z over $0 \leq r \leq 1$ and $0 \leq r \leq 2\pi$. The resulting values are stored in z.

```
>>    [R, Theta] = meshgrid([0:.05:1, 0:pi/20:2*pi]);
>>    Z = z(R, Theta);
```

R, Theta, and Z now contain the ordered triples that are on the surface of $z = r\sin^2(\theta)$ for $0 \leq r \leq 1$ and $0 \leq \theta \leq 2\pi$. Of course, these ordered triples are expressed in cylindrical coordinates.

Step 2

We now need to convert the ordered triples stored in R, Theta, and Z from cylindrical coordinates to rectangular coordinates. This is not hard since we know how to convert the cylindrical coordinate (r,θ,z) to the rectangular coordinate (x,y,z). In particular,

$$x = r\cos(\theta)$$
$$y = r\sin(\theta)$$
$$z = z$$

To carry out this conversion in MATLAB, we first define these conversion functions in MATLAB.

```
>>          x = vectorize(inline(char(r*cos(theta)), 'r', 'theta'));
>>          y = vectorize(inline(char(r*sin(theta)), 'r', 'theta'));
```

Now we can convert the ordered triples in R, Theta, and Z to rectangular coordinates.

```
>>          X = x(R, Theta);
>>          Y = y(R, Theta);
>>          Z = Z;
```

Like R, Theta, and Z, the variables X, Y, and Z now contain the ordered triples that are on the surface of $z = r\sin^2(\theta)$. However, the points in X, Y, and Z are expressed in rectangular coordinates.

Step 3

With the surface for $z = r\sin^2(\theta)$ expressed in rectangular coordinates in X, Y, and Z, we can now plot this surface using MATLAB's surf function.

```
>>          surf(X, Y, Z)
>>          colorbar, xlabel('x'), ylabel('y'), zlabel('z')
```

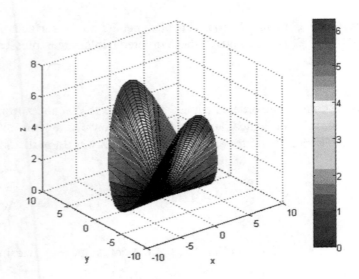

There's no doubt that our lives would be easier if MATLAB provided an `ezcylindrical` function. Of course, we basically just wrote one in this section!

11.2.2 Plotting in Spherical Coordinates

Now let's suppose that we want to plot the following function, which is expressed in spherical coordinates:

$$\rho = 2, \quad 0 \leq \theta \leq \frac{3\pi}{2}, \quad 0 \leq \phi \leq \pi$$

Again, since MATLAB's plotting functions require rectangular coordinates, we must use the following three steps to plot this function in MATLAB:

1. Compute the points on the surface of $z = r\sin^2(\theta)$ in spherical coordinates.
2. Convert these points from spherical coordinates to rectangular coordinates.
3. Use MATLAB to plot the surface defined by these rectangular coordinates.

This is essentially the same process we just used for cylindrical coordinates, so we'll go through the steps more quickly in this section.

Step 1

First, we define the function $\rho = 2$. Since ρ is a constant function, we don't use `char`.

```
>>         syms theta phi
>>         rho = vectorize(inline('2', 'theta', 'phi'));
```

Next, we evaluate ρ over $0 \leq \theta \leq \dfrac{3\pi}{2}$ and $0 \leq \phi \leq \pi$.

```
>>         [Theta, Phi] = meshgrid(0:pi/20:3*pi/2, 0:pi/20:pi);
>>         Rho = rho(Theta, Phi);
```

`Theta`, `Phi`, and `Rho` now contain the ordered triples that are on the surface of $\rho = 2$ for $0 \leq \theta \leq \dfrac{3\pi}{2}$ and $0 \leq \phi \leq \pi$. These ordered triples are expressed in spherical coordinates.

Step 2

We now convert the spherical coordinates in `Theta`, `Phi`, and `Rho` to rectangular coordinates. The equations for converting the spherical coordinate (θ,ϕ,ρ) to the rectangular coordinate (x, y, z) are as follows:

$$x = \rho\sin(\phi)\cos(\theta)$$
$$y = \rho\sin(\phi)\sin(\theta)$$
$$z = \rho\cos(\phi)$$

Thus, we define the following conversion functions in MATLAB.

```
>>         x = vectorize(inline(char(rho(theta,phi)*sin(phi)*
               cos(theta)), 'theta', 'phi'));
>>         y = vectorize(inline(char(rho(theta,phi)*sin(phi)*
               sin(theta)), 'theta', 'phi'));
>>         z = vectorize(inline(char(rho(theta,phi)*cos(phi)),
               'theta', 'phi'));
```

Now we can convert the spherical coordinates in `Theta`, `Phi`, and `Rho` to rectangular coordinates, which we'll store in `X`, `Y`, and `Z`.

```
>>         X = x(Theta, Phi);
>>         Y = y(Theta, Phi);
```

```
>>                  Z = z(Theta, Phi);
```

Step 3

With the surface for $\rho = 2$ expressed in rectangular coordinates in X, Y, and Z, we can plot this surface using MATLAB's `surf` function.

```
>>          surf(X, Y, Z)
>>          view([1, -0.9, 0.2])
>>          colorbar, xlabel('x'), ylabel('y'), zlabel('z')
```

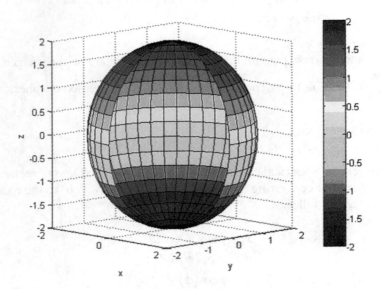

Again, plotting in spherical coordinates would be easier if MATLAB provided an `ezspherical` function. But there's nothing stopping you from writing an `ezspherical` script yourself — that's basically what we just did!

The following functions are expressed in cylindrical coordinates. First, try to picture what each function looks like in your mind. Then, confirm your mental image by plotting the function in MATLAB.

1. $z = 1$, $0 \leq r \leq 1$, $0 \leq \theta \leq 2\pi$
2. $z = 1$, $0.5 \leq r \leq 1$, $0 \leq \theta \leq 2\pi$
3. $z = 1$, $0.5 \leq r \leq 1$, $0 \leq \theta \leq \pi$

4. $z = r$, $\quad 0 \le r \le 1$, $\quad 0 \le \theta \le 2\pi$
5. $z = r$, $\quad 0.5 \le r \le 1$, $\quad 0 \le \theta \le \pi$
6. $z = r$, $\quad 0 \le r \le 1$, $\quad 0.99 \le \theta \le 1$
7. $z = r$, $\quad 0.99 \le r \le 1$, $\quad 0 \le \theta \le 2\pi$
8. $z = r + 1$, $\quad 0 \le r \le 1$, $\quad 0 \le \theta \le 2\pi$
9. $z = r\cos(\theta)$, $\quad 0 \le r \le 1$, $\quad 0 \le \theta \le 2\pi$
10. $z = \sqrt{1 - r^2}$, $\quad 0 \le r \le 1$, $\quad 0 \le \theta \le 2\pi$

The following functions are expressed in spherical coordinates. First, try to picture what each function looks like in your mind. Then confirm your mental image by plotting the function in MATLAB.

1. $\rho = 2$, $\quad 0 \le \theta \le 2\pi$, $\quad 0 \le \phi \le 2\pi$
2. $\rho = 2$, $\quad 0 \le \theta \le \pi$, $\quad 0 \le \phi \le 2\pi$
3. $\rho = 2$, $\quad 0 \le \theta \le 2\pi$, $\quad 0 \le \phi \le \pi$
4. $\rho = 2$, $\quad 0.99 \le \theta \le 1$, $\quad 0 \le \phi \le 2\pi$
5. $\rho = 2$, $\quad 0 \le \theta \le 2\pi$, $\quad 0.99 \le \phi \le 1$
6. $\rho = \sin(\theta)$, $\quad 0 \le \theta \le 2\pi$, $\quad 0 \le \phi \le 2\pi$
7. $\rho = \cos(\theta)$, $\quad 0 \le \theta \le 2\pi$, $\quad 0 \le \phi \le 2\pi$
8. $\rho = \sin(\phi)$, $\quad 0 \le \theta \le 2\pi$, $\quad 0 \le \phi \le 2\pi$
9. $\rho = \cos(\phi)$, $\quad 0 \le \theta \le 2\pi$, $\quad 0 \le \phi \le 2\pi$
10. $\rho = \sec(\phi)$, $\quad 0 \le \theta \le 2\pi$, $\quad 0 \le \phi \le \dfrac{\pi}{4}$

11.3 Conclusion

In this chapter, you learned how to plot in polar, cylindrical, and spherical coordinates in MATLAB. Plotting in polar coordinates turned out to be easy, thanks to MATLAB's ezpolar function. Plotting in cylindrical and spherical coordinates proved to be more difficult, but not overwhelmingly so.